新工科建设之路 · 计算机类专业系列教材

Android 移动应用程序开发

白喆 著

电子工业出版社
Publishing House of Electronics Industry
北京·BEIJING

内 容 简 介

本书通过实例工程介绍如何使用 Java 语言开发 Android 移动应用程序,分为 9 章,内容包括 Android 的基础知识、UI 控件与布局控件、数据适配、基本程序单元、后台服务与广播、数据存储与共享、多媒体与传感器、HTTP 网络通信等。同时,本书针对开发过程中的使用技巧和注意事项,给出了"提示",帮助读者理解书中内容。

本书可作为计算机科学与技术、软件工程、网络工程、数字媒体技术等专业 Android 开发课程的教材,也可供具有 Java 基础的编程爱好者参考。

未经许可,不得以任何方式复制或抄袭本书之部分或全部内容。
版权所有,侵权必究。

图书在版编目(CIP)数据

Android 移动应用程序开发 / 白喆著. —北京:电子工业出版社,2020.7
ISBN 978-7-121-38790-6

I. ①A… II. ①白… III. ①移动终端－应用程序－程序设计－高等学校－教材 IV. ①TN929.53

中国版本图书馆 CIP 数据核字(2020)第 047031 号

责任编辑:章海涛
文字编辑:张 鑫
印　　刷:北京捷迅佳彩印刷有限公司
装　　订:北京捷迅佳彩印刷有限公司
出版发行:电子工业出版社
　　　　　北京市海淀区万寿路 173 信箱　　邮编:100036
开　　本:787×1092　1/16　印张:18.5　字数:474 千字
版　　次:2020 年 7 月第 1 版
印　　次:2024 年 8 月第 9 次印刷
定　　价:62.00 元

凡所购买电子工业出版社图书有缺损问题,请向购买书店调换。若书店售缺,请与本社发行部联系,联系及邮购电话:(010)88254888,88258888。
质量投诉请发邮件至 zlts@phei.com.cn,盗版侵权举报请发邮件至 dbqq@phei.com.cn。
本书咨询联系方式:zhangxinbook@126.com。

前　言

　　Google 的 Android 系统是目前主流的移动设备操作系统之一，具有源代码开源、硬件和开发平台价格低等特点。与 iOS 系统相比，Android 系统对设备兼容性、硬件资源利用率、API 传承性和 App 监管程度的要求较低，而且 Android 系统赋予了开发者和用户更多的权限与选择性。使用 iOS 系统的用户需要通过 App Store 安装 App；而 Android 系统没有指定发布平台，甚至开发者可以在自己的网站上发布 App 的安装包。在 App Store 中发布 App 的审核周期较长，且个人开发者每年需支付一定费用，支付后才可以使用物理设备运行测试，否则只能使用模拟器。在这两个系统上，我都开发过 App，开发难度基本相同，各有优势。

　　2014 年，独立开发完成了一个基于 1 公里半径生活圈的 App——微距（http://www.weiju2014.com），其中包含了社交和购物的功能。我大约用了 10 个月时间完成了 Android 系统的微距 App 和后台网站的程序开发，又用了 2 个月时间完成了 iOS 系统的微距 App 开发，感觉在 iOS 系统开发最"幸福"的地方是无须考虑设备和 API 版本的兼容性，这也是 Android 系统最大的痛点。在微距 App 策划阶段，考虑到对于个人开发者而言，宣传推广和高并发的后台服务器可能是最大的困难，而且积累不了大量的用户就无法发展起来，后来将微距 App 的社交功能分离出来，在此基础上增加了用户筛选和评分的功能，形成了一个免费的交友 App——未见（http://weijian.weiju2014.com）。

　　2018 年，开始准备写一本 Android 开发的零基础教材，从最基础的 Java 和 Kotlin 语法开始写起，一直到综合实例未见 App 开发完成为止。除不包含游戏和系统安全的内容外，涵盖社交、新闻、购物等 App 所包含的知识点，并且搭建一个供学习者练习使用的服务器平台。

　　2019 年偶然的一天，与张鑫编辑沟通时聊起了这个想法，决定写一本精简版，即适合高校教学的 Android 开发教材，这就是这本书的由来。我和张鑫编辑商量后确定了这本教材的基本思路：知识点能够适应高校毕业生第一年工作的需要，并且教学内容能够在一个学期内完成。Android 开发课程通常会作为大二或大三的专业选修课，距离毕业至少有 1 年时间，待学生毕业后，Android 10 会较为普及，因此本书选择目前最新的 API level 29 作为基础，摒弃所有过时的 API，使用 Google 发布的最新标准，不使用向下兼容的方法。同时，考虑到学生毕业后可能会遇到需要向下兼容的情况，本书配套资源提供了几个使用 Google 已经不推荐使用或摒弃的方法的工程文件，感兴趣的学生可以自行查看源代码，书中不再赘述。

　　本书按照 App 使用过程中涉及的各种功能对知识点进行分类，并归纳总结了相关类的常用方法和常量。由于智能手机的普及程度高，所以书中大量使用了高度概括和抽象的方式进行描述，"提示"中介绍相关知识点和使用方法，弥补读者知识储备的不足。本书没有将多线程、异步处理、文件操作、权限、动画、绘制等通用性强且较为容易理解的内容设立为独立的章节，而是将其融入实例工程中。实例工程多以典型的应用场景进行演示，掌握方法后可以举一反三，根据需要灵活应用。Activity、Fragment、Service、IntentService、BroadcastReceiver 等类的实例工程则以流程的方式进行演示。对于所需代码量较大或需要向下兼容的情况，通过"拓展工程"提供给读者，源代码可在本书配套资源中查看。本书实例均使用原生代码编写，没有使用第三方类库或官方扩展类库，因此那些能够自行实现的功能，建议尽量使用原生代码编写。

本书可作为软件工程、计算机科学与技术、网络工程、数字媒体技术等专业 Android 开发课程的教材，也可供具有 Java 基础的编程爱好者参考。

对自学本书的读者建议：第 1 章根据内容逐步操作即可；第 2～9 章先阅读基础知识，了解相关类的常用方法或常量，然后按照步骤完成实例工程，联想使用过的 App 中哪部分使用了实例工程中的技术，再思考实例工程中没有使用到的方法或常量还能实现什么功能，最后将常用方法或常量及实例工程进行一次强化复习。

学习完本书之后，理论上读者已经能够开发具有基本功能的社交类、新闻类、购物分享类、技能分享类、拍照类、录制视频类、音乐播放类 App。当然，App 只有这些基本功能是远远不够的，还需要适合的 UI 界面和后台服务器的支持，服务器端推荐使用 CentOS+Nginx+PHP+ MySQL。如果是小团队或个人开发 App，定位、支付、推送、地图、手机验证码、二维码识别、视频、通信等功能建议直接使用第三方提供的服务，可以节省大量的开发时间，降低技术难度，减少运营成本。

本书配备素材、工程源代码和基础工程源代码，可在华信教育资源网（http：//www.hxedu.com.cn）下载。源代码使用 Android Studio 3.5.3 和 Gradle 5.6.4 进行编写。Android Studio 和 Gradle 更新比较频繁，读者下载时的版本可能更高，打开工程后根据提示更新即可。另外，还提供了 QQ 群（群号：653171771，密码：teachol），加入 QQ 群的读者可以交流学习过程中遇到的问题。

书中的不妥和疏漏之处，欢迎读者批评指正，我的邮箱是 baizhe_22@qq.com，读者可以随时与我联系。

白 喆

2020 年 3 月 31 日

目 录

第1章 Android 的基础知识 1
- 1.1 Android 与 Andy Rubin 1
- 1.2 Android 的开发环境 1
  - 1.2.1 Android Studio 的下载 2
  - 1.2.2 Android Studio 的安装 2
  - 1.2.3 Android SDK 的安装 3
  - 1.2.4 Android Studio 界面 5
  - 1.2.5 Gradle 更新 8
  - 1.2.6 重构 Java 工程 9
- 1.3 创建 Android 工程 10
  - 1.3.1 Android 工程的新建命令 10
  - 1.3.2 Android 工程的创建向导 10
  - 1.3.3 虚拟设备运行工程 13
  - 1.3.4 物理设备运行工程 17
  - 1.3.5 生成签名的 APK 文件 17
- 1.4 Android 的工程结构 19
  - 1.4.1 Project 视图 19
  - 1.4.2 AndroidManifest.xml 文件 19
  - 1.4.3 build.gradle 文件 21
  - 1.4.4 res 文件夹 21
- 1.5 习题 22

第2章 基础 UI 控件 23
- 2.1 UI 控件基础 23
  - 2.1.1 UI 控件的创建方式 23
  - 2.1.2 View 子类的常用属性 24
  - 2.1.3 UI 控件的常用单位 26
- 2.2 文本视图 27
  - 2.2.1 TextView 控件 27
  - 2.2.2 实例工程：显示文本 28
- 2.3 输入框 30
  - 2.3.1 EditText 控件 30

| 2.3.2 | 实例工程：输入发送信息 | 31 |

2.4 按钮 33
2.4.1 Button 控件 33
2.4.2 实例工程：单击按钮获取系统时间 34

2.5 图像视图 35
2.5.1 ImageView 控件 35
2.5.2 实例工程：显示图像 36

2.6 图像按钮 37
2.6.1 ImageButton 控件 37
2.6.2 实例工程：提示广播信息状态的图像按钮 38

2.7 单选按钮 40
2.7.1 RadioButton 控件 40
2.7.2 实例工程：选择性别的单选按钮 41

2.8 复选框 44
2.8.1 CheckBox 控件 44
2.8.2 实例工程：兴趣爱好的复选框 44

2.9 开关按钮 47
2.9.1 Switch 控件 47
2.9.2 实例工程：房间灯的开关按钮 48

2.10 提示信息 50
2.10.1 Toast 控件 50
2.10.2 实例工程：不同位置显示的提示信息 51

2.11 对话框 53
2.11.1 AlertDialog 控件 53
2.11.2 实例工程：默认对话框和自定义对话框 54

2.12 日期选择器 58
2.12.1 DatePicker 控件 58
2.12.2 实例工程：设置日期的日期选择器 59

2.13 时间选择器 61
2.13.1 TimePicker 控件 61
2.13.2 实例工程：设置时间的时间选择器 62

2.14 滚动条视图 64
2.14.1 ScrollView 控件 64
2.14.2 实例工程：滚动显示视图 65

2.15 通知 68
2.15.1 Notification 控件 68

  2.15.2 实例工程：弹出式状态栏通知和自定义视图状态栏通知 ·············· 70
 2.16 习题 ··· 73

第 3 章　UI 布局控件 ··· 74
 3.1 线性布局 ··· 74
  3.1.1 LinearLayout 控件 ·· 74
  3.1.2 实例工程：动态视图的线性布局 ··· 75
 3.2 相对布局 ··· 76
  3.2.1 RelativeLayout 控件 ·· 76
  3.2.2 实例工程：显示方位的相对布局 ··· 77
 3.3 表格布局 ··· 79
  3.3.1 TableLayout 控件 ··· 79
  3.3.2 实例工程：登录界面的表格视图 ··· 80
 3.4 网格布局 ··· 82
  3.4.1 GridLayout 控件 ·· 82
  3.4.2 实例工程：模仿计算器界面的网格布局 ···································· 82
 3.5 帧布局 ··· 84
  3.5.1 FrameLayout 控件 ··· 84
  3.5.2 实例工程：分层显示图像的帧布局 ·· 84
 3.6 约束布局 ··· 85
  3.6.1 ConstraintLayout 控件 ··· 85
  3.6.2 实例工程：模仿朋友圈顶部的约束布局 ···································· 86
 3.7 习题 ··· 87

第 4 章　UI 控件与数据适配 ·· 88
 4.1 数据适配原理 ··· 88
 4.2 列表视图 ··· 89
  4.2.1 ListView 控件 ··· 89
  4.2.2 实例工程：简单数据的列表视图 ··· 90
  4.2.3 实例工程：带缓存的自定义列表视图 ······································· 92
 4.3 网格视图 ··· 96
  4.3.1 GridView 控件 ·· 96
  4.3.2 实例工程：显示商品类别的网格项视图 ···································· 97
 4.4 自动完成文本视图 ·· 101
  4.4.1 AutoCompleteTextView 控件 ·· 101
  4.4.2 实例工程：显示搜索提示的文本框 ··· 102
 4.5 悬浮框 ·· 104
  4.5.1 PopupWindow 控件 ··· 104

4.5.2 实例工程：单击按钮显示自定义悬浮框 ········· 105
4.6 翻转视图 ········· 111
  4.6.1 ViewFlipper 控件 ········· 111
  4.6.2 实例工程：轮流显示图像的翻转视图 ········· 112
4.7 分页视图 ········· 115
  4.7.1 ViewPager 控件 ········· 115
  4.7.2 实例工程：欢迎引导页 ········· 116
4.8 习题 ········· 122

第 5 章 基本程序单元 ········· 123
5.1 活动 ········· 123
  5.1.1 Activity 概述 ········· 123
  5.1.2 Activity 的创建和删除 ········· 125
  5.1.3 Activity 的启动和关闭 ········· 127
  5.1.4 Activity 的生命周期 ········· 129
  5.1.5 Activity 的启动模式 ········· 133
  5.1.6 实例工程：Activity 的数据传递 ········· 138
5.2 碎片 ········· 142
  5.2.1 Fragment 概述 ········· 142
  5.2.2 Fragment 的生命周期 ········· 144
  5.2.3 实例工程：导航分页的主界面 ········· 144
5.3 习题 ········· 150

第 6 章 后台服务与广播 ········· 151
6.1 服务 ········· 151
  6.1.1 Service 概述 ········· 151
  6.1.2 Service 的生命周期 ········· 153
  6.1.3 实例工程：Service 的开启和停止 ········· 153
  6.1.4 实例工程：Service 的绑定和数据传递 ········· 156
  6.1.5 实例工程：Service 显示 Notification ········· 161
6.2 独立线程服务 ········· 163
  6.2.1 IntentService 概述 ········· 163
  6.2.2 实例工程：IntentService 的静态方法启动 ········· 164
6.3 广播接收器 ········· 167
  6.3.1 广播接收器概述 ········· 167
  6.3.2 接收广播 ········· 169
  6.3.3 实例工程：显式和隐式接收广播 ········· 169
  6.3.4 发送广播 ········· 173

		6.3.5 实例工程：发送标准广播和有序广播	173
6.4	习题		176

第 7 章 数据存储与共享 ... 177
- 7.1 共享偏好设置 ... 177
  - 7.1.1 SharedPreferences 概述 ... 177
  - 7.1.2 实例工程：用户登录 ... 179
- 7.2 轻量级数据库 ... 181
  - 7.2.1 SQLite 概述 ... 181
  - 7.2.2 实例工程：自定义通讯录 ... 184
- 7.3 内容提供者 ... 187
  - 7.3.1 ContentProvider 概述 ... 187
  - 7.3.2 实例工程：自定义内容提供者 ... 190
  - 7.3.3 实例工程：访问和修改系统通讯录数据 ... 198
- 7.4 JavaScript 对象表示法 ... 208
  - 7.4.1 JSON 概述 ... 208
  - 7.4.2 实例工程：合成和解析 JSON 数据 ... 210
- 7.5 习题 ... 213

第 8 章 多媒体与传感器 ... 214
- 8.1 系统相机和相册 ... 214
  - 8.1.1 实例工程：拍照、选取和显示图片 ... 214
  - 8.1.2 实例工程：录制、选取和播放视频 ... 221
- 8.2 拍摄照片和录制视频 ... 223
  - 8.2.1 Camera2 类 ... 223
  - 8.2.2 ImageReader 类 ... 230
  - 8.2.3 MediaRecorder 类 ... 231
  - 8.2.4 实例工程：使用 Camera2 类拍摄照片 ... 233
  - 8.2.5 实例工程：使用 Camera2 类录制视频 ... 242
- 8.3 录制音频 ... 249
  - 8.3.1 AudioRecord 类 ... 249
  - 8.3.2 AudioTrack 类 ... 250
  - 8.3.3 实例工程：AudioRecord 录制音频 ... 251
  - 8.3.4 实例工程：MediaRecorder 录制音频 ... 255
- 8.4 传感器 ... 257
  - 8.4.1 传感器概述 ... 257
  - 8.4.2 运动类传感器 ... 259
  - 8.4.3 实例工程：摇一摇比大小 ... 260

		8.4.4 位置类传感器 ·····	262
		8.4.5 实例工程：指南针 ·····	263
		8.4.6 环境类传感器 ·····	266
		8.4.7 实例工程：光照计和气压计 ·····	266
	8.5	位置服务 ·····	268
		8.5.1 位置服务概述 ·····	268
		8.5.2 实例工程：获取经纬度坐标 ·····	270
	8.6	习题 ·····	271
第9章	HTTP 网络通信 ·····		272
	9.1	HttpURLConnection 类 ·····	272
	9.2	实例工程：加载网络图片（带缓存） ·····	273
	9.3	实例工程：发布动态（POST 方式） ·····	278
	9.4	实例工程：动态列表（GET 方式） ·····	282
	9.5	习题 ·····	286

第 1 章　Android 的基础知识

Android 的英文原意是机器人，Android 图标也是一个机器人（如图 1-1 所示），来源于法国作家利尔·亚当在 1886 年发表的科幻小说《未来夏娃》，将外表像人的机器命名为 Android。学习 Android 先从了解 Android 的历史开始，Android 的诞生与 Andy Rubin 有关，Andy Rubin 被称为 Android 之父。

图 1-1　Android 图标

1.1　Android 与 Andy Rubin

1989 年，Andy Rubin 被一名苹果公司工程师引荐到当时处在第一个全盛时期的苹果公司，参与名为 Magic Cap 的智能手机操作系统开发工作。

1999 年，Andy Rubin 创立了 Danger 公司，开发了一个名为 Hiptop 的类似智能手机雏形的设备，提出了"智能手机"的概念——"支持互联网"和"其上运行着能够实现不同功能的各种应用"。

2002 年，Andy Rubin 在斯坦福大学做了一次讲座，听众中包括 Google 公司的两位创始人 Larry Page 和 Sergey Brin。互联网智能手机的理念深深打动了 Larry Page，尤其是他注意到 Danger 产品上默认的搜索引擎为 Google。

2003 年，Andy Rubin 等人创立了 Android 公司，并组建 Android 开发团队，注册了 android.com 域名，立志设计一个基于开源思想的移动平台。当时的手机操作系统都是手机厂商单独开发的，操作系统也是各手机厂商的核心技术，具有很强的封闭性。

2005 年，Andy Rubin 靠自己的积蓄和朋友的支持，艰难地完成了 Android 系统。但在寻找投资方时并不顺利，突然 Andy Rubin 想到了 Google 公司的 Larry Page，于是给他发了一封电子邮件。仅仅几周后，Google 公司收购了成立仅 22 个月的 Android 公司。Andy Rubin 成为 Google 公司工程部副总裁，继续负责 Android 项目。

2014 年，Andy Rubin 离开 Google 公司。2015 年，Andy Rubin 创立 Essential 公司，开发 Android 智能手机。两年后发布首款手机 Essential PH-1，但是销量惨淡。2020 年 2 月 12 日，Essential 公司正式宣布停止运营。

1.2　Android 的开发环境

Android 刚发布时，Google 公司只提供了 SDK，并没有提供官方的开发环境。开发者只能通过第三方工具进行开发，如 Eclipse 和 IntelliJ IDEA。直到 2013 年，Google 公司发布了基于 IntelliJ IDEA 的 Android 集成开发工具——Android Studio，支持 Windows（32 位和 64 位）、Mac 和 Linux 操作系统。

1.2.1 Android Studio 的下载

截至 2020 年 3 月 8 日，Android Studio 的最新版本是 3.5.3，下载网站地址是 https://developer.android.google.cn/studio/。打开网站，单击“DOWNLOAD OPTIONS”链接，跳转至 Android Studio 下载列表（如图 1-2 所示）。

Platform	Android Studio package	Size
Windows (64-bit)	android-studio-ide-192.6241897-windows.exe Recommended	749 MB
	android-studio-ide-192.6241897-windows.zip No .exe installer	752 MB
Windows (32-bit)	android-studio-ide-192.6241897-windows32.zip No .exe installer	751 MB
Mac (64-bit)	android-studio-ide-192.6241897-mac.dmg	762 MB
Linux (64-bit)	android-studio-ide-192.6241897-linux.tar.gz	766 MB
Chrome OS	android-studio-ide-192.6241897-cros.deb	647 MB

图 1-2 Android Studio 下载列表

1.2.2 Android Studio 的安装

1. Windows 版的安装

在 Windows 操作系统中，双击下载的 exe 格式安装文件，根据安装向导的提示安装 Android Studio 和所需的 SDK 工具（如图 1-3 所示）。安装进度完成后，单击“Finish”按钮（如图 1-4 所示）。

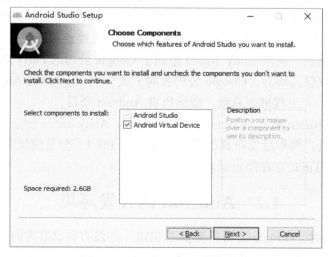

图 1-3 Android Studio 安装向导

图 1-4　Android Studio 完成安装

2．Mac 版的安装

在 Mac 操作系统中，双击下载的 dmg 格式安装文件，在安装界面中将 Android Studio.app 拖曳到 Applications 文件夹中（如图 1-5 所示），即可安装完成。

图 1-5　拖曳安装

1.2.3　Android SDK 的安装

由于目前国内无法直接访问 Android 官方网站的部分地址，在 Windows 操作系统中首次运行时会打开"Android Studio First Run"对话框（如图 1-6 所示）。

图 1-6　"Android Studio First Run"对话框

单击"Setup Proxy"按钮，设置代理地址为"mirrors.neusoft.edu.cn"（如图1-7所示），单击"OK"按钮完成设置。

图1-7 "HTTP Proxy"对话框

> **提示：HTTP Proxy**
> 当无法直接访问Android官方地址进行更新时，需要设置代理地址，否则无法进行自动更新和下载。可以使用东北大学的代理地址：mirrors.neusoft.edu.cn。

在"Android Studio Setup Wizard"对话框中，选择下载的SDK程序和保存路径（如图1-8所示），单击"Next"按钮进行安装，完成安装后单击"Finish"按钮。

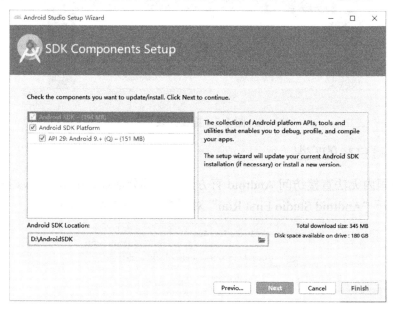

图1-8 Android SDK下载

提示：SDK 占用磁盘空间

Android SDK 占用磁盘空间非常大，至少要准备 10GB 磁盘空间。如果系统盘空间较小，建议安装在其他盘中。

1.2.4 Android Studio 界面

Android Studio 界面主要由菜单栏、工具栏、导航条、左侧工具条、工具窗口、编辑器、右侧工具条、运行工具窗口、状态栏等组成。Mac 版的界面（如图 1-9 所示）与 Windows 版的界面稍有不同。

图 1-9　Mac 版的 Android Studio 界面

1. 菜单栏

菜单栏包含文件（File）、编辑（Edit）、视图（View）、导航（Navigate）、代码（Code）、分析（Analyze）、重构（Refactor）、构建（Build）、运行（Run）、工具（Tools）、版本控制系统（VCS）、窗口（Window）、帮助（Help）等功能菜单（如图 1-10 所示）。

图 1-10　菜单栏

2. 工具栏

工具栏包含从菜单栏中提取出来的一些常用功能（如图 1-11 所示），能够进行快速操作，提高效率。

图 1-11　工具栏

3．导航条

导航条用来辅助查看打开的项目和文件（如图 1-12 所示），单击文件夹可以快速选择子文件夹或文件。

图 1-12　导航条

4．左侧工具条

左侧工具条用来放置窗口的切换按钮（如图 1-13 所示），包含"Project""Resource Manager""Structure""Layout Captures""Build Variants"和"Favorites"窗口。

5．工具窗口

工具窗口最多可以同时显示两个窗口（如图 1-14 所示），单击左侧工具条中的按钮可以进行切换。上面可以显示"Project"或"Resource Manager"窗口，下面可以显示其余的任意一个窗口。

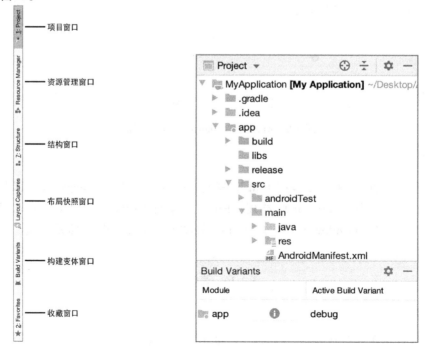

图 1-13　左侧工具条　　　　　图 1-14　工具窗口

6．编辑器

编辑器由文件标签栏、左边栏、编辑区和代码定位栏组成（如图 1-15 所示），是编辑配置信息、编写代码和调试断点设置的区域。

7．右侧工具条

右侧工具条包含"Gradle"窗口和设备文件浏览器（外接 USB 设备或虚拟设备），单击后在编辑区右侧显示（如图 1-16 所示）。

图 1-15　编辑器

图 1-16　右侧工具条

8．运行工具窗口

运行工具窗口主要显示 Android Studio 的运行过程（如图 1-17 所示），左侧显示"Run""TODO""Profiler""Logcat""Terminal"或"Build"窗口，右侧显示"Event Log"窗口。

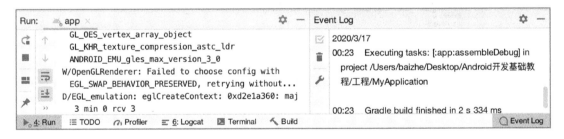

图 1-17　运行工具窗口

9．状态栏

状态栏通常在界面的底部，主要显示 Android Studio 当前的状态和执行的任务（如图 1-18 所示）。

图 1-18　状态栏

1.2.5 Gradle 更新

Gradle 是 Android Studio 默认的 App 构建工具，根据构建规则和配置文件自动构建 App，自动构建包括编译、打包等流程。安装 Android Studio 后，还要注意 Gradle 的更新，Gradle 自动更新时并不通过 Android 官网下载更新文件，而是通过 Gradle 网址（http://services.gradle.org/distributions/）。在国内下载速度较慢甚至无法下载，此时可通过手动下载的方式进行更新。

1．Windows 版的手动更新

将下载的新版本 Gradle 压缩包解压到 gradle 文件夹中（如图 1-19 所示）。

图 1-19　更新版本所在的文件夹

在 Android Studio 中，选择【File】→【Settings】命令（如图 1-20 所示）。在打开的对话框中，选择左侧的"Gradle"选项，然后选择右侧的"Use local gradle distribution"单选按钮，单击"Gradle home"后面的路径选择按钮，选择刚才解压的 gradle 文件夹，单击"OK"按钮（如图 1-21 所示）。

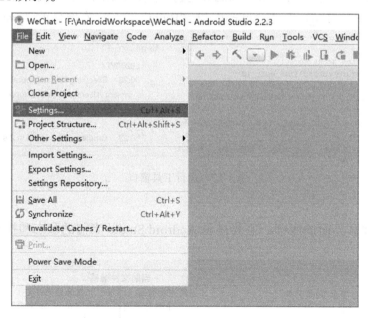

图 1-20　选择【File】→【Settings】命令

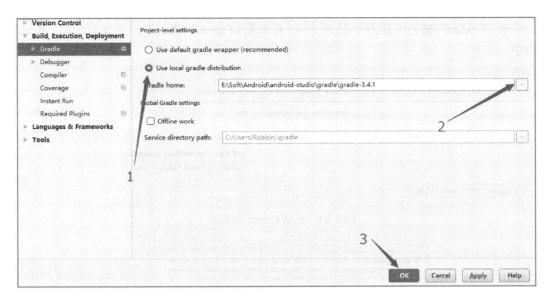

图 1-21　设置本地 Gradle 路径

2．Mac 版的手动更新

在启动台中打开终端，输入"open.gradle"命令后回车，打开更新文件所在的文件夹。将手动下载的"gradle-x.x-all.zip"文件放置在相应的文件夹中（如图 1-22 所示）。重启 Android Studio 后会自动进行更新（如图 1-23 所示）。

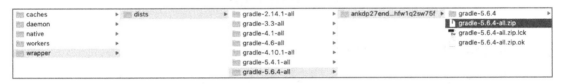

图 1-22　更新文件所在文件夹

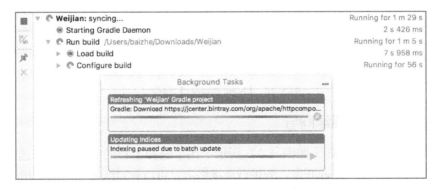

图 1-23　重启 Android Studio 后自动更新 Gradle

1.2.6　重构 Java 工程

Android Studio 升级 Gradle 后，打开原有工程会自动弹出"Plugin Update Recommended"提示框（如图 1-24 所示）。单击后打开"Android Gradle Plugin Update Recommended"对话

框，单击"Update"按钮（如图 1-25 所示），自动重构 Java 工程，详细进度显示在"Build"窗口中。

图 1-24　更新 Gradle 的提示

图 1-25　更新 Gradle 的对话框

1.3　创建 Android 工程

1.3.1　Android 工程的新建命令

选择【File】→【New】→【New Project】命令（如图 1-26 所示），打开"Create New Project"对话框，进行工程创建的向导。

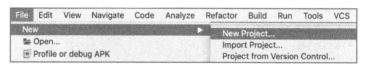

图 1-26　【New Project】命令

1.3.2　Android 工程的创建向导

工程创建向导的第 1 页（如图 1-27 所示）用于选择工程的类型，建议选择"Empty Activity"进行手机项目的 App 开发。若进行单项练习，可以针对练习项目选择相应的类型。

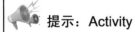

提示：Activity

Activity 是一个应用组件，用户可与其进行交互，用于绘制用户界面的窗口。窗口通常会充满屏幕，也可小于屏幕并浮动在其他窗口之上。后续将详细介绍 Activity。

单击"Next"按钮，进入工程创建向导的第 2 页（如图 1-28 所示），包含"Name""Package name""Save location""Language""Minimum SDK"等配置选项。

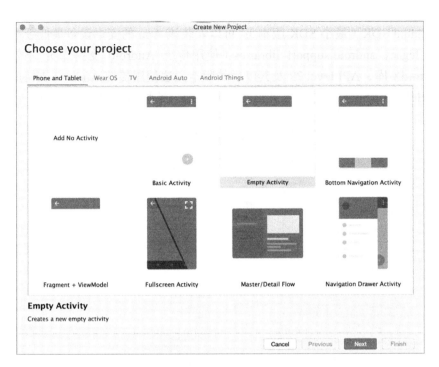

图 1-27　选择工程类型

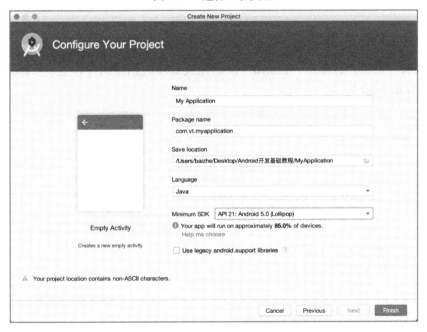

图 1-28　配置工程

- Name（应用名称）：安装到 Android 设备上后显示的名称。
- Package name（包名）：一般以企业域名倒置作为前缀，然后加上 App 的英文名。
- Save location（存储位置）：保存工程文件的路径。
- Language（语言）：可以选择 Java 或 Kotlin。

- Minimum SDK（最小 SDK 版本）：可以安装该 App 的安卓设备最低版本。
- Use legacy android.support libraries（使用传统 Android 支持库）：不勾选时使用 AndroidX 库，API level 29 及其以后的版本无法使用传统 Android 支持库。

单击"Help me choose"选项，可查看官方统计的各版本使用比例（如图 1-29 所示）。

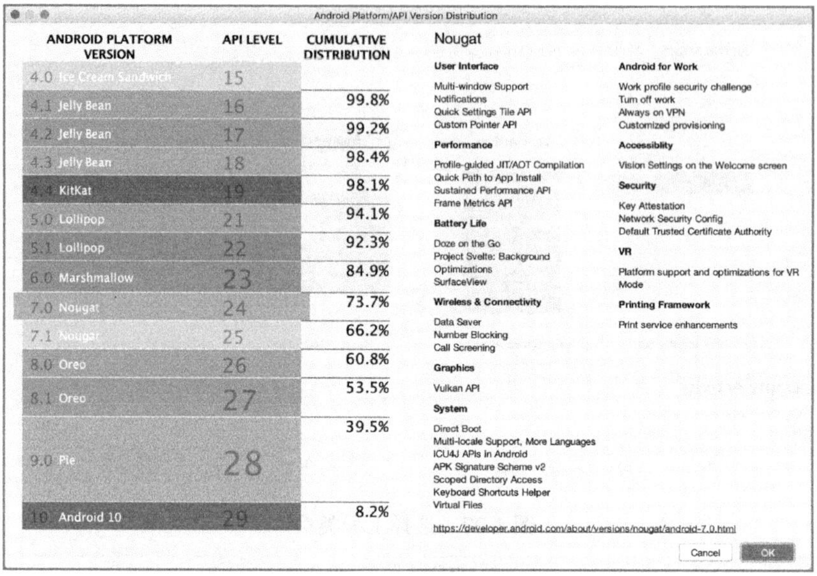

图 1-29　各版本使用比例

单击"OK"按钮，再单击"Finish"按钮完成工程的创建向导，显示 Android Studio 界面（如图 1-30 所示）。

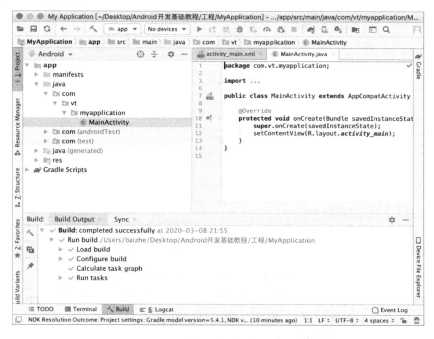

图 1-30　创建工程完成后的 Android Studio 界面

1.3.3 虚拟设备运行工程

1. 创建虚拟设备

在工具栏中，单击"AVD Manager"按钮（如图 1-31 所示），打开"Android Virtual Device Manager"对话框。

图 1-31 "AVD Manager"按钮

在"Android Virtual Device Manager"对话框中，单击"Create Virtual Device"按钮（如图 1-32 所示），打开"Virtual Device Configuration"对话框。

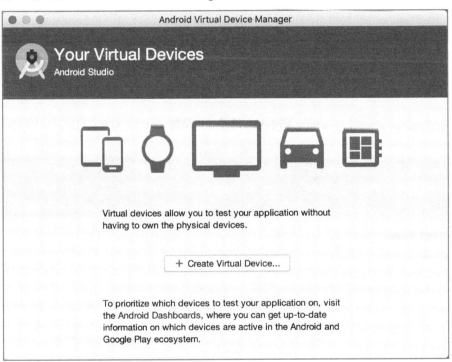

图 1-32 "Android Virtual Device Manager"对话框

在"Virtual Device Configuration"对话框中，显示虚拟设备的参数列表（如图 1-33 所示）。根据需求选择相应的虚拟设备，然后单击"Next"按钮。

在选择 Android 系统的版本时，如果没有下载相应的系统镜像，需要选择"DownLoad"选项下载系统镜像（如图 1-34 所示）。

选择"DownLoad"选项后，会下载 SDK 文件并自动安装。安装完成后，单击"Finish"按钮，确认虚拟设备的配置信息（如图 1-35 所示），然后单击"Next"按钮完成虚拟设备的创建。

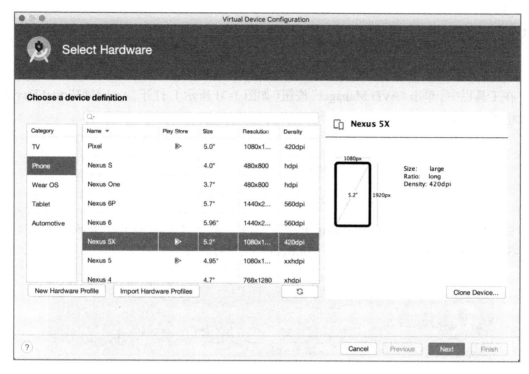

图 1-33 选择虚拟设备

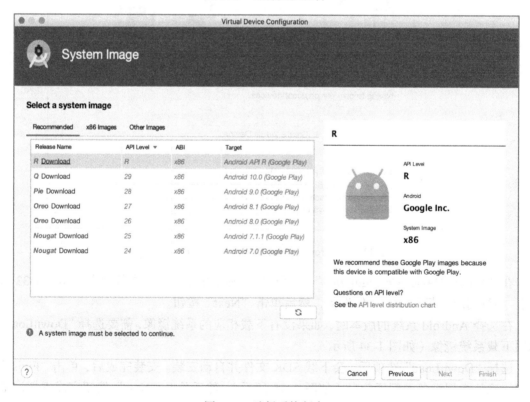

图 1-34 选择系统版本

第 1 章 Android 的基础知识

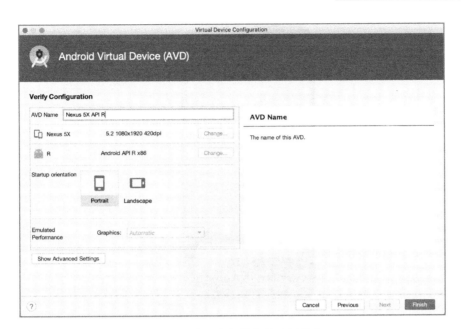

图 1-35　确认虚拟设备的配置信息

返回"Android Virtual Device Manager"对话框，选择相应的虚拟设备，单击"Launch this AVD in the emulator"图标（如图 1-36 所示），启动虚拟设备。

图 1-36　启动虚拟设备

Android 虚拟设备启动时间较长，若硬件配置较好，则可以大大缩短启动时间。启动完成后，显示虚拟设备界面（如图 1-37 所示）。

15

图 1-37　虚拟设备界面

2. 运行工程

在工具栏中，选择创建的虚拟设备，然后单击"Run 'app'(^R)"按钮（如图 1-38 所示）。工程编译完成后，自动安装在虚拟设备中，然后自动启动运行（如图 1-39 所示）。

图 1-38　"Run 'app'(^R)"按钮

图 1-39　使用虚拟设备运行

1.3.4 物理设备运行工程

使用 USB 线将手机连接到计算机，在工具栏的运行/调试设备下拉菜单中会显示连接的手机作为首选项（如图 1-40 所示）。单击"Run 'app'(^R)"按钮，会编译工程并安装在连接的手机上，然后自动启动运行（如图 1-41 所示）。

图 1-40　显示连接的手机　　　　　图 1-41　手机运行画面

1.3.5 生成签名的 APK 文件

APK 文件是 App 的安装文件，设置 APK 文件签名的目的是让 App 不被恶意生成的 APK 文件覆盖安装。在升级 App 时，只有同一签名的 APK 文件才能对 App 进行升级，从而避免恶意覆盖。

选择【Build】→【Generate Signed Bundle/APK】命令，打开"Generate Signed Bundle or APK"对话框，选择"APK"单选按钮，单击"Next"按钮（如图 1-42 所示）进入下一个页面。

图 1-42　选择"APK"单选按钮

单击"Create new"按钮，打开"New Key Store"对话框，设置签名文件密码、签名密码和证书内容（如图 1-43 所示），然后单击"OK"按钮完成签名文件的创建。

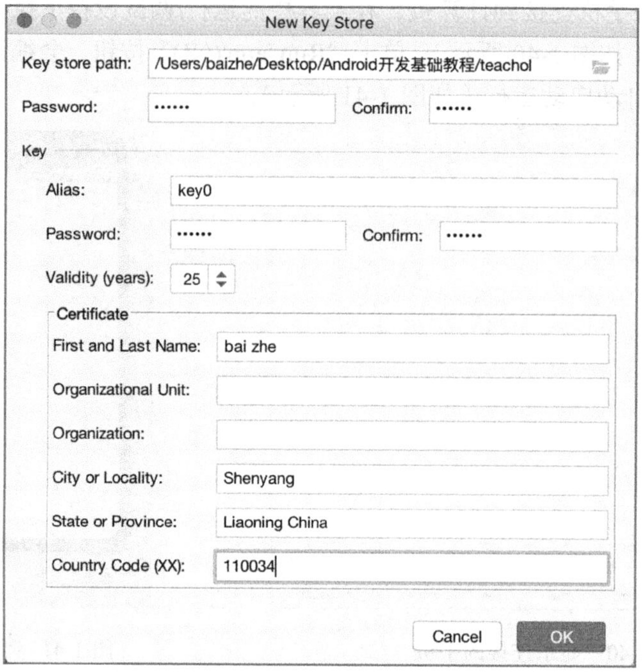

图 1-43 "New Key Store"对话框

创建签名文件后，在"Generate Signed Bundle or APK"对话框中会自动输入签名路径、签名路径密码、签名别名和签名密码（如图 1-44 所示），单击"Next"按钮，进入下一个页面。"Build Variants"选择"release"，勾选"Signature Versions"中的"V2（Full APK Signature）"复选框（如图 1-45 所示），单击"Finish"按钮生成签名的 APK 文件。

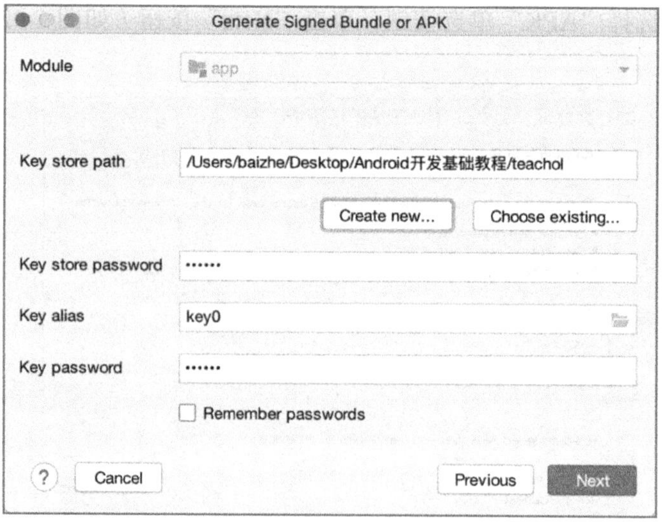

图 1-44 自动输入签名文件信息

第 1 章　Android 的基础知识

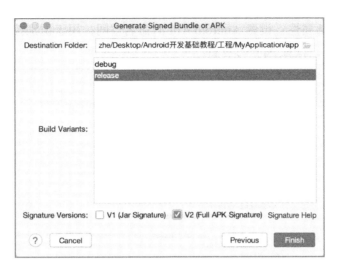

图 1-45　选择完整版和完整签名

> **提示：防反编译**
> 对 APK 文件进行反编译可以还原源代码，为了防止源代码被恶意使用，商业上使用的 App 还会进行防反编译处理。常用的防反编译方法包括混淆策略、整体 Dex 加固、拆分 Dex 加固、虚拟机加固等，推荐使用阿里聚安全、腾讯云应用乐固、爱加密、娜迦、梆梆等自动化加固处理以防反编译。

1.4　Android 的工程结构

1.4.1　Project 视图

Project 视图是以工程管理的形式显示文件的视图（如图 1-46 所示），主要包括 AndroidManifest.xml 文件（App 的配置信息）、build.gradle 文件（gradle 配置文件）、res 文件夹（资源文件）和 java 文件夹（源代码和测试代码）。

1.4.2　AndroidManifest.xml 文件

AndroidManifest.xml（如图 1-47 所示）是 Android 应用的配置文件，包含应用的基本信息，声明程序中的 Activity、ContentProvider、Service 和 Receiver，指定 permission 和 instrumentation。

1．<manifest>标签

<manifest>标签是 AndroidManifest.xml 的根节点，用于设置 xmlns:android 和 package 属性，还包括一个<application>标签。

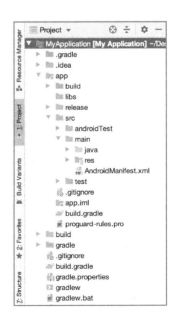

图 1-46　Project 视图

19

- xmlns:android 属性：设置 Android API 的命名空间，用于识别控件的属性。
- package 属性：设置 App 的包名。

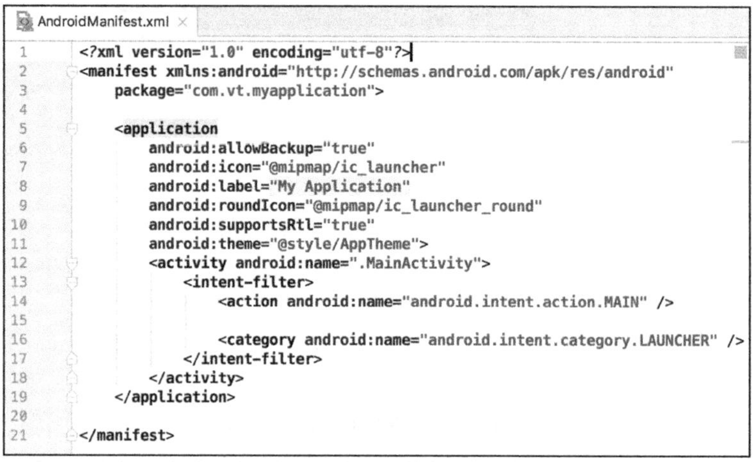

```
<?xml version="1.0" encoding="utf-8"?>
<manifest xmlns:android="http://schemas.android.com/apk/res/android"
    package="com.vt.myapplication">

    <application
        android:allowBackup="true"
        android:icon="@mipmap/ic_launcher"
        android:label="My Application"
        android:roundIcon="@mipmap/ic_launcher_round"
        android:supportsRtl="true"
        android:theme="@style/AppTheme">
        <activity android:name=".MainActivity">
            <intent-filter>
                <action android:name="android.intent.action.MAIN" />

                <category android:name="android.intent.category.LAUNCHER" />
            </intent-filter>
        </activity>
    </application>

</manifest>
```

图 1-47　AndroidManifest.xml 文件

2．<application>标签

<application>标签用于声明应用程序，包含每个应用程序组件所声明的子元素，以及能够影响所有组件的属性。

- android:allowBackup 属性：设置 App 数据的备份和恢复功能，默认值为 true，可以通过 adb backup 和 adb restore 对 App 数据进行备份和恢复。
- android:icon 属性：设置一个 Drawable 资源作为 App 的普通图标（如图 1-48 所示）。
- android:label 属性：设置 App 的名称。
- android:roundIcon 属性：设置一个 Drawable 资源作为 App 的圆形图标（如图 1-49 所示）。

图 1-48　默认普通图标

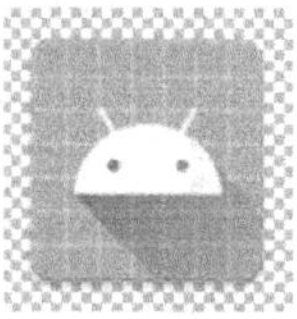

图 1-49　默认圆形图标

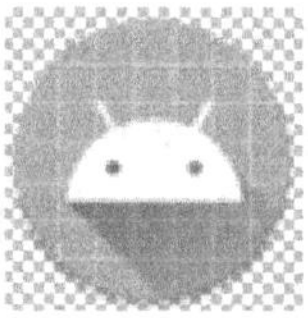

- android:supportsRtl 属性：设置是否支持从右到左的布局，默认值为 true。当 build.gradle 文件中的 targetSdkVersion 值设置为 17 或更高时，可以使用 RTL（right-to-left）布局。
- android:theme 属性：设置界面皮肤主题。

3．<activity>标签

<activity>标签用于设置 Activity 的属性，Activity 必须被声明在<manifest>标签中，没有被声明的 Activity 不会被调用。android:name 属性设置声明的 Activity 的类名。

4. <intent-filter>标签

<intent-filter>标签用于指定 Activity、Service 或 BroadcastReceiver 能够响应的 Intent 对象类型，通过<action>、<category>和<data>子标签进行描述。

5. <action>标签

<action>标签用于给 Intent 过滤器添加子标签，<intent-filter>标签必须包含一个或多个 <action>标签。如果不包含<action>标签，就不存在 Intent 过滤器。当 android:name 属性设置为 android.intent.action.MAIN 时，该 Activity 作为应用默认启动的 Activity。

6. <category>标签

<category>标签用于向 Intent 过滤器添加类别名称，android:name 属性的默认值为 CATEGORY_DEFAULT。

1.4.3 build.gradle 文件

本工程中包含 2 个 build.gradle 文件，分别是根目录下的顶级（top level）build.gradle 文件和 app 目录下的模块级（module level）build.gradle 文件。

顶级 build.gradle 文件位于项目根目录，用于配置适合项目中所有模块的构建设置。默认情况下，顶级 build.gradle 文件使用 buildscript{}代码块来定义项目中所有模块共用的 Gradle 存储区和依赖项。

模块级 build.gradle 文件位于 project/module 文件夹中，用于配置适合其所在模块的构建设置，通过配置这些构建设置可以提供自定义打包选项。

1.4.4 res 文件夹

Android 的资源文件存储在工程的 res 文件夹中（如图 1-50 所示），Android Studio 3 将资源分类存储在 4 类文件夹（前缀名为 drawable、layout、mipmap 和 values）中。drawable-v24 文件夹的后缀 v24 表示 API 的版本，mipmap-mdpi 文件夹的后缀 mdpi 表示 Android 设备的屏幕分辨率。

- drawable：用于存储图像和 XMLDrawable 文件，XMLDrawable 分为 AnimationDrawable、BitmapDrawable、ClipDrawable、ColorDrawable、GradientDrawable、NinePatchDrawable、RotateDrawable、StateListDrawable、ShapeDrawable 和 TransitionDrawable 等类型。
- layout：用于存储界面布局文件。
- mipmap：用于存储图标文件，默认包含自动生

图 1-50 res 文件夹结构

成的 Launcher Icons（桌面图标），还可以存储 Action Bar and Tab Icons（桌面图标）和 Notification Icons（通知图标）。
- values：用于存储 colors.xml、strings.xml 和 styles.xml 等文件。

1.5 习　　题

1. 下载最新版本的 Android Studio，安装后如果需要更新 Gradle，则将其更新到最新版本。
2. 创建一个任意类型的手机 App 工程，分别使用虚拟设备和物理设备运行。
3. 在 build.gradle 文件中修改支持的 API 最低版本，然后生成带签名的 APK 文件。

第 2 章　基础 UI 控件

用户最直接接触 App 的部分是 App 的界面，App 界面的核心组成部分是 UI 控件。Android 提供了丰富的 UI 控件，不但能呈现丰富多彩的内容，还能进行人机交互。基础 UI 控件可以单独使用，无须依靠其他 UI 控件或适配器就可以实现核心功能。

2.1　UI 控件基础

2.1.1　UI 控件的创建方式

UI（User Interface）控件提供了界面元素和布局模型的支持。View 子类的 UI 控件，可以使用两种方式创建：XML 标签创建和动态创建。非 View 子类的 UI 控件只能通过动态创建。

XML 标签创建是指在 XML 布局文件中使用 XML 标签创建 UI 控件并设置各种属性，在与其关联的基本程序单元中呈现出来；动态创建是指在基本程序单元中使用代码动态创建 UI 控件，并设置各种属性。

1．XML 标签创建 UI 控件

新建"Empty Activity"工程后，打开"activity_main.xml"文件，可以直接预览界面效果。将左侧"Palette"中的控件拖曳到中间的预览视图中，选择控件后，可以在右侧"Attributes"窗口中设置属性（如图 2-1 所示）。

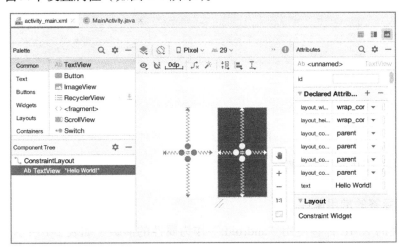

图 2-1　XML 布局文件的预览

单击右上角的"Code"按钮直接查看 UI 控件标签的代码（如图 2-2 所示），可以通过代码编辑布局文件。

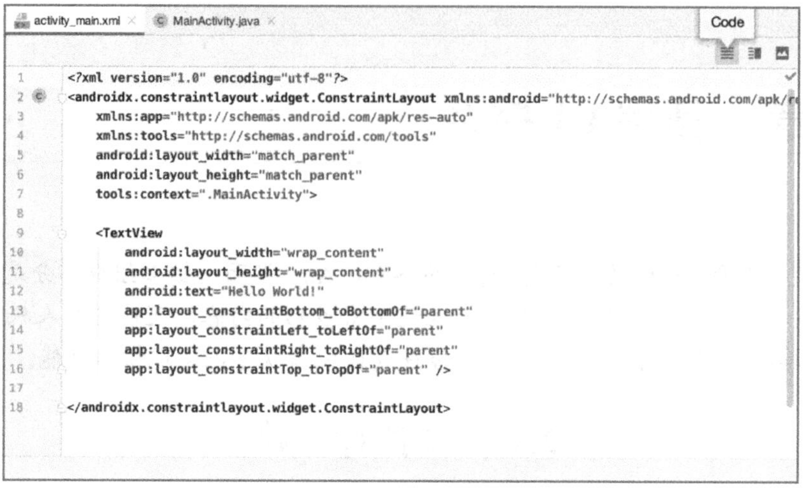

图 2-2　XML 布局文件的代码

2．动态创建 UI 控件

新建"Empty Activity"工程后，打开"MainActivity.java"文件，使用 Java 代码创建一个 TextView 控件的实例，然后使用 setContentView() 方法将其添加到视图中并显示出来（如图 2-3 所示）。

图 2-3　动态创建 UI 控件

2.1.2　View 子类的常用属性

1．XML 标签的常用属性

View 子类的 XML 标签使用"android:"作为属性的前缀（如表 2-1 所示），使用"="设置属性。

表 2-1　XML 标签的常用属性

属　性	说　　明
android:id	设置标识符
android:layout_width	设置宽度，属性值包括： ● 具体的长度值，长度单位推荐使用 dp ● wrap_content，自动匹配所包含的子控件宽度 ● match_parent，填充父控件，表示使当前控件的大小和父布局的大小一样 ● fill_parent，与 match_parent 的作用相同，已经不推荐使用
android:layout_height	设置高度，属性值同 android:layout_width
android:visibility	设置可见性，属性值包括： ● visible，可见 ● invisible，不可见，但占据原来的屏幕空间 ● gone，不可见，也不会占据屏幕空间
android:background	设置背景
android:foreground	设置前景
android:alpha	设置透明度（0~1 之间的数值，表示百分比）
android:padding	设置内边距（推荐使用 dp 作为单位）
android:layout_margin	设置外边距（推荐使用 dp 作为单位）
android:scrollbars	设置在滚动时是否显示滚动条

2．常用的 View 类方法

View 子类设置 UI 控件的属性，不但可以设置 XML 标签提供的属性，还提供了获取属性及更多功能的方法（如表 2-2 所示）。

表 2-2　View 子类设置 UI 控件属性的常用方法

类型和修饰符	方　　法
int	getId() 获取标识符
Object	getTag() 获取标记对象
void	setId(int id) 设置标识符
void	setTag(Object tag) 设置与视图关联的标记对象
void	setLayoutParams(ViewGroup.LayoutParams params) 设置布局参数
void	setVisibility(int visibility) 设置可见性
void	setBackground(Drawable background) 设置背景
void	setAlpha(float alpha) 设置透明度
void	setPadding(int left, int top, int right, int bottom) 设置内边距，单位为像素
void	setFocusable(boolean focusable) 设置是否可以获取焦点

续表

类型和修饰符	方法
final void	requestFocus() 获取焦点
void	clearFocus() 清除焦点
void	setOnClickListener(View.OnClickListener l) 设置单击监听器
void	setOnLongClickListener(View.OnLongClickListener l) 设置长时间单击监听器
void	setOnFocusChangeListener(View.OnFocusChangeListener l) 设置改变焦点监听器
void	setOnScrollChangeListener(View.OnScrollChangeListener l) 设置滚动改变监听器
void	setOnTouchListener(View.OnTouchListener l) 设置触碰监听器

提示：类方法设置和获取标签属性

控件的标签属性多数可以使用 setXXX()方法设置，使用 getXXX()方法获取，布尔值的属性使用 isXXX()方法获取。例如，android:id 属性可以使用 setId()方法设置，使用 getId()方法获取。

2.1.3　UI 控件的常用单位

UI 控件的常用单位有 px、dp 和 sp，此外还可以使用 mm、in、pt 等。不同 Android 设备屏幕的 dpi（像素密度）不同（如表 2-3 所示），dp 和 sp 单位可以解决不同 dpi 下显示的差异问题，在显示前会自动根据 dpi 转换成相应的 px 单位值。Android 项目中的控件大小主要使用 dp 作为单位，字体大小主要使用 sp 作为单位。

表 2-3　屏幕像素密度

参　　数	LDPI	MDPI	HDPI	XHDPI	XXHPDI	XXXHDPI
像素密度	120dpi	160dpi	240dpi	320dpi	480dpi	640dpi
分辨率(像素)	240×320	320×480	480×800	720×1280	1080×1920	3840×2160
转换系数	0.75	1	1.5	2	3	4
转换结果	1dp=0.75px	1dp=1px	1dp=1.5px	1pd=2px	1dp=3px	1dp=4px

- px：像素（pixel），每个单位像素代表屏幕上的一个显示点。100px 的图像在不同分辨率的手机上显示的大小是不同的，即使同样尺寸的屏幕，分辨率也可能不同，因此不建议使用该单位。
- dp：设备独立像素，等同于 dip（density-independent pixel），转换公式为 dp × dpi/160=px。160dpi 的中密度手机屏幕为基准屏幕，此时 1dp=1px。100dp 在 320×480（MDPI，160dpi）的手机上是 100px，100dp 在 480×800（HDPI，240dpi）的手机上是 150px，但它们都是 100dp。不管屏幕的像素密度是多少，相同 dp 大小的元素在屏幕上显示的大小始终都差不多，因此控件多使用该单位。

- sp：与缩放无关的抽象像素（scale-independent pixel），与 dp 类似，转换公式为 sp × dpi/160=px。当用户通过手机设置修改手机字体时，以 sp 为单位的字体会随着改变，因此字号多使用该单位。

> 提示：分辨率、屏幕大小、dpi 和 ppi
> - 分辨率：手机屏幕的像素点数，一般描述成屏幕的"宽×高"。常见的分辨率有 480 像素×800 像素、720 像素×1280 像素、1080 像素×1920 像素等。
> - 屏幕大小：屏幕大小是手机对角线的物理尺寸，以英寸（inch）为单位。5 英寸手机是指对角线长 5 英寸，5 英寸×2.54 厘米/英寸=12.7 厘米。
> - dpi（dot per inch）：每英寸多少点，该值越高，图像越细腻。
> - ppi（pixel per inch）：每英寸多少像素，该值越高，屏幕显示越细腻。

2.2 文本视图

2.2.1 TextView 控件

TextView 控件（android.widget.TextView）是用于显示文本信息的 UI 控件，是 android.view.View 的子类。TextView 控件的常用 XML 标签属性如表 2-4 所示，TextView 类的常用方法如表 2-5 所示。

表 2-4 TextView 控件的常用 XML 标签属性

属 性	说 明
android:text	设置文本内容
android:gravity	设置文本的对齐方式，可选值有 top、bottom、left、right、center_vertical、center_horizontal、center 等，可以用竖线连接使用多个值
android:textSize	设置文本大小
android:textColor	设置文本颜色

表 2-5 TextView 类的常用方法

类型和修饰符	方 法
void	setGravity(int gravity) 设置对齐方式
final void	setText(CharSequence text) 设置文本内容
void	setTextSize(float size) 设置文本大小，单位为 sp
void	setTextColor(@ColorInt int color) 设置文本颜色
CharSequence	getText() 获取文本内容
int	length() 获取文本长度
void	addTextChangedListener(TextWatcher watcher) 添加文本改变监听器

提示：UI 控件标签和类

在程序运行中，标签设置的属性只能通过类的实例进行关联后修改，而类不但通过对应的方法设置标签属性，还提供了其他操作类的方法。由于篇幅限制，在后续 UI 控件内容中，只提供标签形式的常用属性表，在类的常用方法表中不提供设置和获取属性值的方法。

2.2.2 实例工程：显示文本

本实例演示了显示文本的两种方法（如图 2-4 所示）。第一个文本使用标签创建，使用 setText()方法修改文本内容；第二个文本通过代码动态创建，使用代码设置内容、字号、布局等属性。

1．新建工程

新建一个"Empty Activity"工程，工程名称为"C0201"。

2．主界面的布局

打开"activity_main.xml"文件，将原有标签删除，重新添加标签如下。

图 2-4　运行效果

```
/res/layout/activity_main.xml
01  <?xml version="1.0" encoding="utf-8"?>
02  <LinearLayout xmlns:android="http://schemas.android.com/apk/res/android"
03      android:id="@+id/linear_layout_main_activity"
04      android:layout_width="match_parent"
05      android:layout_height="match_parent"
06      android:orientation="vertical">
07      <TextView
08          android:id="@+id/text_view_hello"
09          android:layout_width="match_parent"
10          android:layout_height="match_parent"
11          android:textSize="36sp"
12          android:text="Hello World!" />
13  </LinearLayout>
```

第 02～06 行<LinearLayout>标签是用于线性布局的根标签，android:orientation 属性设置为垂直布局。android:id 属性值为"@+id/linear_layout_main_activity"，解析后等同于"R.id.linear_layout_main_activity"，供 Java 代码调用。第 07～12 行<TextView>标签用于显示文本，android:text 属性设置显示的文本内容（预览效果如图 2-5 所示）。

第 2 章 基础 UI 控件

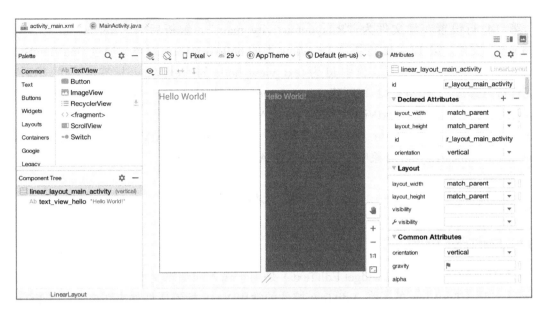

图 2-5 "activity_main.xml" 文件预览效果

3. 主界面的 Activity

```
/java/com/vt/c0201/MainActivity.java
10  public class MainActivity extends AppCompatActivity {
11      @Override
12      protected void onCreate(Bundle savedInstanceState) {
13          super.onCreate(savedInstanceState);
14          setContentView(R.layout.activity_main);//设置布局文件
15          //设置 LayoutParams
16          LinearLayout.LayoutParams helloLayoutParams = new LinearLayout.LayoutParams
                  (LinearLayout.LayoutParams.MATCH_PARENT, 500);
17          //通过 id 获取 TextView 对象并修改属性
18          TextView helloTextView = this.findViewById(R.id.text_view_hello);
19          helloTextView.setGravity(Gravity.CENTER_HORIZONTAL | Gravity.CENTER_VERTICAL);
20          helloTextView.setText("Hello Android!");
21          helloTextView.setTextSize(36f);
22          helloTextView.setLayoutParams(helloLayoutParams);
23          //通过 id 获取 XML 布局中的 linearLayout 布局
24          LinearLayout mainActivityLinearLayout = this.findViewById(R.id.linear_layout_
                  main_activity);
25          //动态创建 TextView 对象
26          TextView welcomeTextView = new TextView(this);
27          welcomeTextView.setGravity(Gravity.CENTER);
28          welcomeTextView.setText("welcome!");
29          welcomeTextView.setTextSize(36f);
30          welcomeTextView.setTextColor(Color.rgb(255, 255, 255));
31          //添加到 linearLayout 布局中
32          mainActivityLinearLayout.addView(welcomeTextView);
33      }
34  }
```

第 14 行设置布局文件为"R.layout.activity_main",解析后为"/res/layout/activity_main.xml"文件。第 18～22 行新建 TextView 类的 helloTextView 对象,通过 findViewById(R.id.text_view_hello)与布局文件中 id 为 text_view_hello 的<TextView>标签相关联,然后对其属性进行修改。第 24～32 行新建 LinearLayout 类的 mainActivityLinearLayout 对象和 TextView 类的 welcomeTextView 对象,mainActivityLinearLayout 对象与布局文件中的<LinearLayout>标签进行关联,然后将 welcomeTextView 对象添加到 mainActivityLinearLayout 对象中,以便显示在 Activity 的视图中。

2.3 输　入　框

2.3.1 EditText 控件

EditText 控件(android.widget.EditText)是用于输入和编辑文本内容的 UI 控件,是 android.widget.TextView 的子类。EditText 控件的常用 XML 标签属性如表 2-6 所示,EditText 类的常用方法如表 2-7 所示。

表 2-6　EditText 控件的常用 XML 标签属性

属性	说明
android:hint	设置提示性的文字,输入内容后提示性的文字消失
android:editable	设置是否可以编辑
android:ems	设置以 em 为单位的控件宽度。em 是一个在印刷排版中使用的单位,表示字宽,可理解为 equal M(和 M 字符一样的宽度为一个单位)
android:focusable	设置是否可以获取焦点
android:maxEms	设置控件的最长宽度为多少个字符的宽度。与 ems 同时使用时覆盖 ems 选项
android:maxLength	设置文本最大长度
android:minLines	设置最小显示行数
android:maxLines	设置最大显示行数
android:selectAllOnFocus	设置获得焦点后是否全选组件内所有文本内容
android:inputType	设置限制输入文本类型,常用的输入类型值包括: ● text,输入类型为普通文本 ● number,输入类型为数字文本 ● phone,输入类型为电话号码 ● numberDecimal,输入类型为小数数字,允许十进制小数点提供分数值 ● numberPassword,输入类型为数字密码 ● textEmailAddress,输入一个电子邮件地址 ● textPassword,输入一个密码
android:singleLine	设置是否单行

表 2-7　EditText 类的常用方法

类型和修饰符	方法
CharSequence	getText() 获取文本内容
void	selectAll() 全选

类型和修饰符	方法
void	setSelection(int start, int stop) 设置选择范围
void	setSelection(int index) 设置光标位置

2.3.2 实例工程：输入发送信息

本实例演示了三种不同输入框的使用方式（如图2-6所示）。在三个输入框中输入内容，第一个输入框输入的字符超过ems属性设置的长度后，之前输入的字符会被推后隐藏起来；第二个输入框输入的字符超过maxEms属性设置的长度后，会在maxLength属性设置的范围内自动延长控件的长度；第三个输入框输入换行后，会在maxLines属性设置的范围内自动增加一行。

图2-6　运行效果

1．新建工程

新建一个"Empty Activity"工程，工程名称为"C0202"。

2．输入框背景的资源文件

在"Project"窗口中切换到"Android"下拉选项，选择"drawable"文件夹，单击右键并选择【New】→【Drawable Resource File】命令（如图2-7所示）。

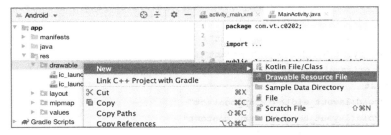

图2-7　选择【New】→【Drawable Resource File】命令

在打开的"New Resource File"对话框中，输入文件名称和根标签名称（如图 2-8 所示），单击"OK"按钮，生成一个资源文件作为多行输入框的背景。

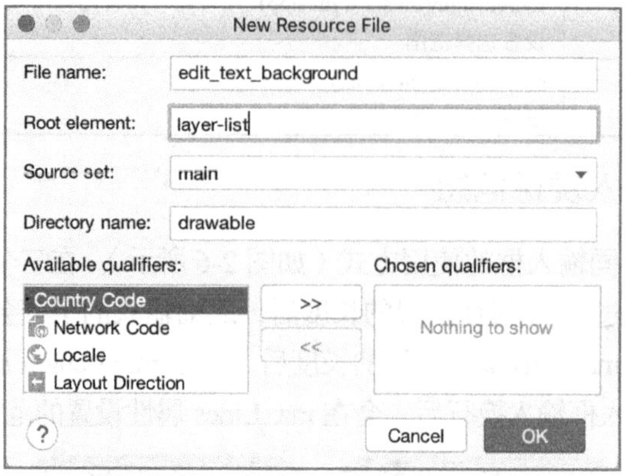

图 2-8 "New Resource File"对话框

```
/res/drawable/edit_text_background.xml
01  <?xml version="1.0" encoding="utf-8"?>
02  <layer-list xmlns:android="http://schemas.android.com/apk/res/android">
03      <item>
04          <shape android:shape="rectangle">
05              <solid android:color="#EFEFEF" />
06              <corners android:radius="3dp" />
07              <stroke android:width="0.5dp" android:color="#505050" />
08          </shape>
09      </item>
10  </layer-list>
```

第 02 行<layer-list>标签是图层列表标签，其原理是一层层叠加图层，后添加的会覆盖先添加的标签。第 04 行<shape>标签设置外形，可以设置为矩形（rectangle）、椭圆形（oval）、线形（line）、环形（ring）。第 05 行<solid>标签设置内部填充颜色。第 06 行<corners>标签设置圆角效果，radius 属性设置圆角的半径。第 07 行<stroke>标签设置描边效果，width 属性设置描边的宽度，color 属性设置描边的颜色。

3. 主界面的布局

```
/res/layout/activity_main.xml
01  <?xml version="1.0" encoding="utf-8"?>
02  <LinearLayout xmlns:android="http://schemas.android.com/apk/res/android"
03      android:id="@+id/linearLayout"
04      android:layout_width="match_parent"
05      android:layout_height="match_parent"
06      android:orientation="vertical">
07      <EditText
08          android:layout_width="wrap_content"
09          android:layout_height="wrap_content"
10          android:layout_marginLeft="10dp"
11          android:layout_marginRight="10dp"
```

```
12          android:ems="4"
13          android:hint="用户名"
14          android:inputType="textPersonName"
15          android:singleLine="true" />
16      <EditText
17          android:layout_width="wrap_content"
18          android:layout_height="wrap_content"
19          android:layout_marginLeft="10dp"
20          android:layout_marginRight="10dp"
21          android:hint="密码"
22          android:inputType="textPassword"
23          android:maxEms="4"
24          android:maxLength="7"
25          android:singleLine="true" />
26      <EditText
27          android:layout_width="match_parent"
28          android:layout_height="wrap_content"
29          android:layout_margin="10dp"
30          android:background="@drawable/edit_text_background"
31          android:hint="需要发送的信息"
32          android:maxLines="5"
33          android:padding="5dp" />
34  </LinearLayout>
```

第 07~15 行是第一个输入框的标签,android:inputType="textPersonName"表示输入类型为个人名字,android:singleLine="true"表示单行。第 16~25 行是第二个输入框的标签,android:inputType="textPassword"表示输入类型为密码,输入后会以星号替代显示。第 26~33 行是第三个输入框的标签,android:background="@drawable/edit_text_background"设置背景,"@drawable/edit_text_background" 解析后对应的文件为 "/res/drawable/edit_text_background.xml",并将该文件中的设置作为背景。android:maxLines="5"表示最多可以输入 5 行。

2.4 按 钮

2.4.1 Button 控件

Button 控件(android.widget.Button)是按钮的 UI 控件,是 android.widget.TextView 的子类。Button 控件的常用 XML 标签属性如表 2-8 所示。

表 2-8 Button 控件的常用 XML 标签属性

属 性	说 明
android:text	设置按钮上的文本
android:textColor	设置按钮上的文本颜色
android:textSize	设置按钮上的文本字号
android:enabled	设置按钮是否可用
android:gravity	设置按钮上文本的对齐方式

2.4.2 实例工程：单击按钮获取系统时间

本实例演示了按钮及其单击监听事件的应用，单击"获取当前系统时间"按钮，在其上方的 TextView 控件中显示当前的系统时间（如图 2-9 所示）。

图 2-9　运行效果

1．新建工程

新建一个"Empty Activity"工程，工程名称为"C0203"。

2．主界面的布局

```
/res/layout/activity_main.xml
01  <?xml version="1.0" encoding="utf-8"?>
02  <LinearLayout xmlns:android="http://schemas.android.com/apk/res/android"
03      android:layout_width="match_parent"
04      android:layout_height="match_parent"
05      android:gravity="center"
06      android:orientation="vertical">
07      <TextView
08          android:id="@+id/text_view_time"
09          android:layout_width="wrap_content"
10          android:layout_height="wrap_content" />
11      <Button
12          android:id="@+id/button_get_time"
13          android:layout_width="wrap_content"
14          android:layout_height="wrap_content"
15          android:text="获取当前系统时间" />
16  </LinearLayout>
```

第 07～10 行<TextView>标签用于显示获取的系统时间，第 11～15 行<Button>标签用于单击后获取当前系统时间并显示在<TextView>标签中。

3．主界面的 Activity

```
/java/com/vt/c0203/MainActivity.java
11  public class MainActivity extends AppCompatActivity {
12      @Override
13      protected void onCreate(Bundle savedInstanceState) {
14          super.onCreate(savedInstanceState);
15          setContentView(R.layout.activity_main);
16          final TextView textView = this.findViewById(R.id.text_view_time);
17          Button button = this.findViewById(R.id.button_get_time);
18          //添加监听器
19          button.setOnClickListener(new View.OnClickListener() {
20              @Override
21              public void onClick(View v) {
22                  Date date = new Date();
23                  SimpleDateFormat dateFormat = new SimpleDateFormat("yyyy年MM月dd日HH时mm分ss秒");
24                  textView.setText(dateFormat.format(date));
25              }
26          });
27      }
28  }
```

第 16 行使用 final 关键字修饰与 activity_main.xml 中<TextView>标签相关联的 TextView 对象，被 final 修饰的局部变量可以用在内联方法中。第 19～26 行 button 对象与 activity_main.xml 中的<Button>标签相关联，并使用内联类创建一个单击事件监听器，用于获取系统时间并显示在 TextView 对象中。

2.5 图 像 视 图

2.5.1 ImageView 控件

ImageView 控件（android.widget.ImageView）是用于显示图像的 UI 控件，是 android.view.View 的子类。图像通常存储在"drawable"资源文件夹中或下载后显示出来。ImageView 控件的常用 XML 标签属性如表 2-9 所示，ImageView 类的常用方法如表 2-10 所示。

表 2-9 ImageView 控件的常用 XML 标签属性

属　　性	说　　明
android:adjustViewBounds	设置是否保持比例调整视图边界，默认值是 false
android:maxHeight	设置最大高度
android:maxWidth	设置最大宽度
android:src	设置图像的路径
android:scaleType	设置图像显示的缩放匹配类型，类型包含：fitCenter（默认值）、center、centerCrop、centerInside、fitEnd、fitStart、fitXY 和 matrix
android:background	设置背景

表 2-10　ImageView 类的常用方法

类型和修饰符	方　　法
void	setImageAlpha(int alpha) 设置图像透明度
void	setSelected(boolean selected) 设置是否被选择

2.5.2　实例工程：显示图像

本实例演示了显示图像及其三种缩放匹配效果（如图 2-10 所示）。第一张图像是等比例最大化居中显示的；第二张图像是以原始尺寸在左上角显示的；第三张图像是完全填充显示的，可能会导致图像被拉伸变形。

图 2-10　运行效果

1．新建工程并导入素材

新建一个"Empty Activity"工程，工程名称为"C0204"。然后选择"/res/mipmap"资源文件夹，将素材图像（素材文件夹路径为"/素材/C0204"）粘贴或拖曳到该文件夹中（如图 2-11 所示）。

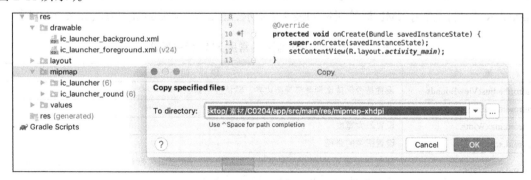

图 2-11　添加图像

2．主界面的布局

```
/res/layout/activity_main.xml
01  <?xml version="1.0" encoding="utf-8"?>
02  <LinearLayout xmlns:android="http://schemas.android.com/apk/res/android"
03      android:layout_width="match_parent"
04      android:layout_height="match_parent"
05      android:gravity="center_horizontal"
06      android:orientation="vertical">
07      <ImageView
08          android:id="@+id/image_view1"
09          android:layout_width="180dp"
10          android:layout_height="150dp"
11          android:layout_margin="5dp"
12          android:background="#00BCD4"
13          android:padding="3dp"
14          android:src="@mipmap/img01" />
15      <ImageView
16          android:id="@+id/image_view2"
17          android:layout_width="180dp"
18          android:layout_height="150dp"
19          android:layout_margin="5dp"
20          android:background="#00BCD4"
21          android:padding="3dp"
22          android:scaleType="matrix"
23          android:src="@mipmap/img02" />
24      <ImageView
25          android:id="@+id/image_view3"
26          android:layout_width="180dp"
27          android:layout_height="150dp"
28          android:layout_margin="5dp"
29          android:background="#00BCD4"
30          android:padding="3dp"
31          android:scaleType="fitXY"
32          android:src="@mipmap/img03" />
33  </LinearLayout>
```

3个ImageView控件分别显示不同的图像，设置了相同的背景颜色和内间距。第一个没有使用scaleType属性值，另外两个的scaleType属性值分别为"matrix"和"fitXY"。

2.6 图像按钮

2.6.1 ImageButton 控件

ImageButton控件（android.widget.ImageButton）是图像按钮的UI控件，是android.widget.ImageView的子类，不但能够显示图像，而且能实现按钮的功能；除了继承的方法，还包含两个公有方法（如表2-11所示）。

表 2-11　ImageButton 类的公有方法

类　　型	方　　法
CharSequence	getAccessibilityClassName() 获取此对象的类名以用于辅助功能
PointerIcon	onResolvePointerIcon(MotionEvent event, int pointerIndex) 获取运动事件的指针图标，如果未指定该图标，则返回空

2.6.2　实例工程：提示广播信息状态的图像按钮

本实例演示了单击图像按钮，图像中喇叭上方小圆点消失，再次单击，小圆点出现的效果（如图 2-12 所示）。

1. 新建工程导入素材

新建一个"Empty Activity"工程，工程名称为"C0205"。然后选择"/res/drawable"资源文件夹，将素材图像（素材文件夹路径为"/素材/C0205"）粘贴或拖曳到该文件夹中。

2. 图像按钮选择状态的资源文件

在"/res/drawable"资源文件夹上单击右键，选择【New】→【Drawable resource file】命令（如图 2-13 所示）。在打开的"New Resource File"对话框的"File name"文本框中输入"image_button_msg"，单击"OK"按钮，创建"image_button_msg.xml"文件。该文件用于设置 ImageButton 控件不同状态时显示的图像资源。

图 2-12　运行效果

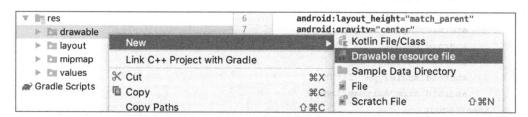

图 2-13　新建 Drawable 资源文件

```
/res/drawable/image_button_msg.xml
01  <?xml version="1.0" encoding="utf-8"?>
02  <selector xmlns:android="http://schemas.android.com/apk/res/android">
03      <item android:drawable="@mipmap/img_msg_new" android:state_selected="false" />
04      <item android:drawable="@mipmap/img_msg" android:state_selected="true" />
05  </selector>
```

第 03 行设置 ImageButton 控件未被选择时显示的图像，@mipmap/img_msg_new 表示 mipmap 文件夹中的 img_msg_new.png。第 04 行设置 ImageButton 控件被选择时显示的图像。

提示:StateListDrawable 资源状态列表

在资源文件的<selector>标签下的<item>标签中,可以设置资源不同状态显示效果的属性,与 "android:drawable" 属性搭配使用。该属性值为布尔型,常用的状态属性如下。
- android:state_selected: 表示控件是否被选择的状态。
- android:state_pressed: 表示控件是否被单击或触摸的状态。
- android:state_focused: 表示控件是否获得焦点时的状态。
- android:state_checkable: 表示控件是否可以被勾选的状态。
- android:state_checked: 表示控件是否处于被勾选的状态。
- android:state_enabled: 表示控件是否处于可用的状态。
- android:state_activated: 表示控件是否被激活的状态。

3. 主界面的布局

```
/res/layout/activity_main.xml
01  <?xml version="1.0" encoding="utf-8"?>
02  <LinearLayout xmlns:android="http://schemas.android.com/apk/res/android"
03      android:layout_width="match_parent"
04      android:layout_height="match_parent"
05      android:gravity="center">
06      <ImageButton
07          android:id="@+id/image_button"
08          android:layout_width="100dp"
09          android:layout_height="100dp"
10          android:background="#E62F5A"
11          android:src="@mipmap/image_button_msg" />
12  </LinearLayout>
```

第 10 行设置背景颜色,当图像有透明部分或超出图像尺寸时,会显示为背景颜色。第 11 行未直接设置显示的图像资源,而是通过资源配置文件设置不同状态时显示的图像。

4. 主界面的 Activity

```
/java/com/vt/c0205/MainActivity.java
09  public class MainActivity extends AppCompatActivity {
10      @Override
11      protected void onCreate(Bundle savedInstanceState) {
12          super.onCreate(savedInstanceState);
13          setContentView(R.layout.activity_main);
14          ImageButton button = this.findViewById(R.id.image_button);
15          button.setOnClickListener(new View.OnClickListener() {
16              @Override
17              public void onClick(View v) {
18                  if (v.isSelected()) {
19                      v.setSelected(false);
20                      Log.i("ImageButton", "选中图像");
21                  } else {
22                      v.setSelected(true);
23                      Log.i("ImageButton", "取消选中图像");
```

```
24                }
25            }
26        });
27    }
28 }
```

第 14 行将 ImageButton 对象与 activity_main.xml 中的<ImageButton>标签相关联。第 15~26 行使用内联类创建单击事件监听器,用于切换图像按钮的显示状态。onClick(View v) 方法的 v 参数表示被单击的 View 对象,由于 ImageButton 类是 View 类的子类,因此可以直接传递给 v 参数。第 20 行和第 23 行的 Log.i()在控制台中输出测试信息,与 system.out.println()相比,主要优势在于能使用过滤器过滤日志。

5. 测试运行

选择【Run】→【Run'app'(^R)】命令后,在"Logcat"窗口中单击过滤器的输入框后,可以选择或输入"ImageButton"作为输出信息的过滤标签,用于筛选 Log.i()输出的信息(如图 2-14 所示)。

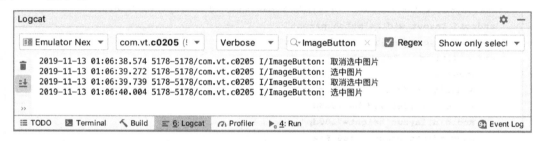

图 2-14 "Logcat"窗口筛选信息

 **提示:Log 类**

Log 类直接调用静态方法输出调试信息,在"Logcat"窗口中输出,并可以通过过滤器对<tag>标签进行筛选。"Logcat"窗口中输出的调试信息有以下 5 种形式。
- Verbose:任何信息都会输出,使用 Log.v()。
- Info:输出提示信息,使用 Log.i()。
- Error:输出错误信息,使用 Log.e()。
- Debug:输出调试信息,使用 Log.d()。
- Warning:输出警告信息,使用 Log.w()。

2.7 单选按钮

2.7.1 RadioButton 控件

RadioButton 控件(android.widget.RadioButton)是单选按钮的 UI 控件,是 android.widget. CompoundButton 的子类。只有同一个 RadioButton 中的 RadioButton 才能实现真正的单选功能。RadioButton 控件的常用 XML 标签属性如表 2-12 所示,RadioButton 类的常用方法

如表 2-13 所示；RadioButton 控件的常用 XML 标签属性如表 2-14 所示，RadioButton 类的常用方法如表 2-15 所示。

表 2-12　RadioButton 控件的常用 XML 标签属性

属　　性	说　　明
android:button	设置文本前方的按钮形状，为@null 时不显示前面的圆形按钮
android:checked	设置是否被选中，对同一个 RadioButtonGroup 中的 RadioButton 设置该属性后会取消之前被选中的 RadioButton 的选中状态
android:text	设置文本

表 2-13　RadioButton 类的常用方法

类型和修饰符	方　　法
boolean	isChecked() 判断单选按钮是否被选择
void	setChecked(boolean checked) 设置单选按钮是否被选择
void	setOnCheckedChangeListener(CompoundButton.OnCheckedChangeListener listener) 设置单选按钮选择状态改变的监听器

表 2-14　RadioButton 控件的常用 XML 标签属性

属　　性	说　　明
android:checkedButton	设置默认选择的 RadioButton 选项 id
android:orientation	设置 RadioButton 的排列方式。"horizontal" 为水平排列，"vertical" 为垂直排列

表 2-15　RadioButton 类的常用方法

类型和修饰符	方　　法
void	addView(View child, int index, ViewGroup.LayoutParams params) 添加子视图
void	check(int id) 设置被选择的单选按钮
void	clearCheck() 清除选择
int	getCheckedRadioButtonId() 获取被选择单选按钮的 id
void	setOnCheckedChangeListener(RadioGroup.OnCheckedChangeListener listener) 设置单选按钮组选择状态改变的监听器

2.7.2　实例工程：选择性别的单选按钮

本实例演示了使用单选按钮选择性别（如图 2-15 所示），单击改变选择的选项或提交按钮时，会在控制台中输出相应的信息。

1．新建工程

新建一个"Empty Activity"工程，工程名称为"C0206"。

图 2-15 运行效果

2. 字符串资源

/res/values/strings.xml	
01	`<resources>`
02	`    <string name="App_name">C0206</string>`
03	`    <string name="select_sex">请选择性别：</string>`
04	`    <string name="man">男</string>`
05	`    <string name="woman">女</string>`
06	`    <string name="submit">提交</string>`
07	`</resources>`

第 03～06 行是添加的字符串资源，用于设置控件的 text 属性。与直接设置 text 属性相比，其优势在于可以重复调用和国际化，便于统一修改和针对不同语言设置字符串。

3. 主界面的布局

/res/layout/activity_main.xml	
01	`<?xml version="1.0" encoding="utf-8"?>`
02	`<LinearLayout xmlns:android="http://schemas.android.com/apk/res/android"`
03	`    android:layout_width="match_parent"`
04	`    android:layout_height="match_parent"`
05	`    android:orientation="vertical">`
06	`    <TextView`
07	`        android:layout_width="wrap_content"`
08	`        android:layout_height="wrap_content"`
09	`        android:text="@string/select_sex" />`
10	`    <RadioGroup`
11	`        android:id="@+id/radio_group"`
12	`        android:layout_width="wrap_content"`
13	`        android:layout_height="wrap_content"`
14	`        android:checkedButton="@id/radio_button_women"`
15	`        android:orientation="horizontal">`
16	`        <RadioButton`
17	`            android:id="@+id/radio_button_man"`

```
18            android:layout_width="wrap_content"
19            android:layout_height="wrap_content"
20            android:text="@string/man" />
21        <RadioButton
22            android:id="@+id/radio_button_women"
23            android:layout_width="wrap_content"
24            android:layout_height="wrap_content"
25            android:text="@string/woman" />
26    </RadioGroup>
27    <Button
28        android:id="@+id/button"
29        android:layout_width="wrap_content"
30        android:layout_height="wrap_content"
31        android:text="@string/submit" />
32 </LinearLayout>
```

第 10～15 行<RadioGroup>标签设置默认的 RadioButton 选项和 RadioButton 的排列方式。第 16～25 行添加两个<RadioButton>标签。如果在<RadioButton>标签中设置了 checked 属性，将覆盖<RadioGroup>标签中设置的 checkedButton 属性及同组中之前<RadioButton>标签已经设置的 checked 属性。

4．主界面的 Activity

```
/java/com/vt/c0206/MainActivity.java
11 public class MainActivity extends AppCompatActivity {
12     @Override
13     protected void onCreate(Bundle savedInstanceState) {
14         super.onCreate(savedInstanceState);
15         setContentView(R.layout.activity_main);
16         final RadioGroup radioGroup = findViewById(R.id.radio_group);
17         radioGroup.setOnCheckedChangeListener(new RadioGroup.OnCheckedChangeListener() {
18             @Override
19             public void onCheckedChanged(RadioGroup radioGroup, int checkedId) {
20                 RadioButton radioButton = findViewById(checkedId);
21                 Log.i("RadioButton", "您当前选择的选项: " + radioButton.getText());
22             }
23         });
24         Button button = findViewById(R.id.button);
25         button.setOnClickListener(new View.OnClickListener() {
26             @Override
27             public void onClick(View v) {
28                 for (int i = 0; i < radioGroup.getChildCount(); i++) {
29                     RadioButton radioButton = (RadioButton) radioGroup.getChildAt(i);
30                     if (radioButton.isChecked()) {
31                         Log.v("RadioButton", "您提交的选项是:" + radioButton.getText());
32                         break;
33                     }
34                 }
35             }
36         });
37     }
38 }
```

第 17～23 行添加 RadioGroup 对象改变选项的监听事件，用于在控制台中输出改变选项后的信息。第 25~36 行添加 Button 对象的单击监听事件，用于获取选中的 RadioButton 对象，并在控制台中输出。

2.8 复 选 框

2.8.1 CheckBox 控件

CheckBox 控件（android.widget.Checkbox）是复选框的 UI 控件，是 android.widget.CompoundButton 的子类。CheckBox 控件的常用 XML 标签属性如表 2-16 所示，CheckBox 类的常用方法如表 2-17 所示。

表 2-16 CheckBox 控件的常用 XML 标签属性

属性	说明
android:button	设置文本前方的按钮形状，为@null 时不显示前面的方形按钮
android:checked	设置是否被选中
android:text	设置文本

表 2-17 CheckBox 类的常用方法

类型和修饰符	方法
boolean	isChecked() 判断复选框是否被选择
void	setChecked(boolean checked) 设置复选框是否被选择
void	setOnCheckedChangeListener(CompoundButton.OnCheckedChangeListener listener) 设置复选框选择状态改变的监听器

2.8.2 实例工程：兴趣爱好的复选框

本实例演示了使用复选框选择兴趣爱好（如图 2-16 所示）。未选择时，单击"确定"按钮提示未选择兴趣爱好；选择后，单击"确定"按钮显示所选的兴趣爱好。

图 2-16 运行效果

1. 新建工程

新建一个"Empty Activity"工程,工程名称为"C0207"。

2. 主界面的布局

```
/res/layout/activity_main.xml
01  <?xml version="1.0" encoding="utf-8"?>
02  <LinearLayout xmlns:android="http://schemas.android.com/apk/res/android"
03      android:layout_width="match_parent"
04      android:layout_height="match_parent"
05      android:orientation="vertical">
06      <TextView
07          android:layout_width="match_parent"
08          android:layout_height="wrap_content"
09          android:text="请选择你的兴趣爱好: " />
10      <CheckBox
11          android:id="@+id/check_box1"
12          android:layout_width="wrap_content"
13          android:layout_height="wrap_content"
14          android:text="读书" />
15      <CheckBox
16          android:id="@+id/check_box2"
17          android:layout_width="wrap_content"
18          android:layout_height="wrap_content"
19          android:text="旅行" />
20      <CheckBox
21          android:id="@+id/check_box3"
22          android:layout_width="wrap_content"
23          android:layout_height="wrap_content"
24          android:text="摄影" />
25      <CheckBox
26          android:id="@+id/check_box4"
27          android:layout_width="wrap_content"
28          android:layout_height="wrap_content"
29          android:text="绘画" />
30      <Button
31          android:id="@+id/button_submit"
32          android:layout_width="match_parent"
33          android:layout_height="wrap_content"
34          android:text="确定" />
35      <TextView
36          android:id="@+id/text_view_result"
37          android:layout_width="match_parent"
38          android:layout_height="wrap_content" />
39  </LinearLayout>
```

第 10～29 行添加了 4 个 CheckBox 控件,用于选择兴趣爱好。第 30～34 行添加了一个 Button 控件,用于确定选择。第 35～38 行添加了一个 TextView 控件,用于显示选择结果。

3. 主界面的 Activity

```
/java/com/vt/c0207/MainActivity.java
12   public class MainActivity extends AppCompatActivity implements CompoundButton.OnChecked
     ChangeListener {
13       private ArrayList<String> mHobbies = new ArrayList<>();
14       @Override
15       protected void onCreate(Bundle savedInstanceState) {
16           super.onCreate(savedInstanceState);
17           setContentView(R.layout.activity_main);
18
19           final TextView resultTextView = findViewById(R.id.text_view_result);
20           CheckBox checkBox1 = findViewById(R.id.check_box1);
21           CheckBox checkBox2 = findViewById(R.id.check_box2);
22           CheckBox checkBox3 = findViewById(R.id.check_box3);
23           CheckBox checkBox4 = findViewById(R.id.check_box4);
24           Button submitButton = findViewById(R.id.button_submit);
25           //CheckBox 设置监听器
26           checkBox1.setOnCheckedChangeListener(this);
27           checkBox2.setOnCheckedChangeListener(this);
28           checkBox3.setOnCheckedChangeListener(this);
29           checkBox4.setOnCheckedChangeListener(this);
30           //Button 设置监听器
31           submitButton.setOnClickListener(new View.OnClickListener() {
32               @Override
33               public void onClick(View v) {
34                   StringBuilder sb = new StringBuilder();
35                   for (int i = 0; i < mHobbies.size(); i++) {
36                       //选择的兴趣爱好添加到 StringBuilder 尾部
37                       if (i == (mHobbies.size() - 1)) {
38                           sb.append(mHobbies.get(i));
39                       } else {
40                           sb.append(mHobbies.get(i) + "、");
41                       }
42                   }
43                   //显示选择结果
44                   if (sb.length() == 0) {
45                       resultTextView.setText("您还没有进行选择了!");
46                   } else {
47                       resultTextView.setText("您选择了: " + sb + "。");
48                   }
49               }
50           });
51       }
52       /**
53        * 实现改变选项的接口方法
54        * @param compoundButton
55        * @param isChecked
56        * @return void
57        */
```

```
58      @Override
59      public void onCheckedChanged(CompoundButton compoundButton, boolean isChecked) {
60          if (isChecked) {
61              //添加到数组
62              mHobbies.add(compoundButton.getText().toString().trim());
63          } else {
64              //从数组中移除
65              mHobbies.remove(compoundButton.getText().toString().trim());
66          }
67      }
68  }
```

第 12 行继承 CompoundButton.OnCheckedChangeListener 接口，用于实现 CheckBox 类的改变选项状态的监听事件接口。第 13 行定义一个 ArrayList 动态数组，用于保存选择的兴趣爱好。第 25 行添加一行注释，用于注解说明下方的代码。第 52~57 行添加文档注释，@param 表示方法的参数，@return 表示方法的返回值。

> **提示：注释**
> 注释用于说明某段代码的作用或说明某个类的用途、某个方法的功能，以及该方法的参数、返回值的数据类型和意义等。书写注释的方式有以下 3 种。
> - 单行注释：最常用的注释方式，注释内容从"//"开始，到本行末尾结束。
> - 多行注释：注释内容从"/*"开始，到"*/"结束。
> - 文档注释：注释内容从"/**"开始，到"*/"结束。专门用于生成帮助文档的注释，可以包含以@开头的预设标签。

2.9 开关按钮

2.9.1 Switch 控件

Switch 控件（android.widget.Switch）是开关按钮的 UI 控件，是 android.widget.CompoundButton 的子类，Switch 控件的常用 XML 标签属性如表 2-18 所示，Switch 类的常用方法如表 2-19 所示。

表 2-18 Switch 控件的常用 XML 标签属性

属性	说明
android:checked	设置 on/off 的状态
android:showText	设置 on/off 时是否显示文本
android:switchMinWidth	设置开关的最小宽度
android:textOff	设置 off 状态时显示的文本
android:textOn	设置 on 状态时显示的文本
android:track	设置底部的图像
android:thumb	设置滑块的图像

表 2-19　Switch 类的常用方法

类型和修饰符	方　　法
boolean	isChecked() 判断开关按钮是否打开
void	setChecked(boolean checked) 设置开关按钮是否打开
void	setOnCheckedChangeListener(CompoundButton.OnCheckedChangeListener listener) 设置开关按钮状态改变的监听器
void	toggle() 改变选择状态

2.9.2　实例工程：房间灯的开关按钮

本实例演示了两种效果的开关按钮（如图 2-17 所示）。单击 Switch 控件后，监听事件会判断被单击的控件及其状态，并在"Logcat"窗口中输出相应的信息。

图 2-17　运行效果

1．新建工程

新建一个"Empty Activity"工程，工程名称为"C0208"。

2．主界面的布局

```
/res/layout/activity_main.xml
01  <?xml version="1.0" encoding="utf-8"?>
02  <LinearLayout xmlns:android="http://schemas.android.com/apk/res/android"
03      android:layout_width="match_parent"
04      android:layout_height="match_parent"
05      android:gravity="center"
06      android:orientation="vertical">
07      <Switch
```

08	`        android:id="@+id/switch_livingroom"`
09	`        android:layout_width="wrap_content"`
10	`        android:layout_height="wrap_content"`
11	`        android:showText="true"`
12	`        android:switchMinWidth="50dp"`
13	`        android:text="客厅灯"`
14	`        android:textOff="关"`
15	`        android:textOn="开" />`
16	`    <Switch`
17	`        android:id="@+id/switch_bedroom"`
18	`        android:layout_width="wrap_content"`
19	`        android:layout_height="wrap_content"`
20	`        android:checked="true"`
21	`        android:text="卧室灯" />`
22	`</LinearLayout>`

第 07～21 行添加两个<Switch>标签,第一个标签设置显示开关的文本及不同状态的文本,第二个标签设置选中的状态。

3. 主界面的 Activity

	`/java/com/vt/c0208/MainActivity.java`
09	`public class MainActivity extends AppCompatActivity implements CompoundButton.OnCheckedChangeListener {`
10	`    @Override`
11	`    protected void onCreate(Bundle savedInstanceState) {`
12	`        super.onCreate(savedInstanceState);`
13	`        setContentView(R.layout.activity_main);`
14	`        Switch livingroomSwitch = findViewById(R.id.switch_livingroom);`
15	`        Switch bedroomSwitch = findViewById(R.id.switch_bedroom);`
16	`        livingroomSwitch.setOnCheckedChangeListener(this);`
17	`        bedroomSwitch.setOnCheckedChangeListener(this);`
18	`    }`
19	`    //重写开关状态改变的事件`
20	`    @Override`
21	`    public void onCheckedChanged(CompoundButton compoundButton, boolean isChecked) {`
22	`        switch (compoundButton.getId()) {`
23	`            case R.id.switch_livingroom:`
24	`                if (isChecked) {`
25	`                    Log.i("Switch", "打开客厅灯");`
26	`                } else {`
27	`                    Log.i("Switch", "关闭客厅灯");`
28	`                }`
29	`                break;`
30	`            case R.id.switch_bedroom:`
31	`                if (isChecked) {`
32	`                    Log.i("Switch", "打开卧室灯");`
33	`                } else {`
34	`                    Log.i("Switch", "关闭卧室灯");`
35	`                }`
36	`                break;`

37	}
38	}
39	}

第 09 行 implements CompoundButton.OnCheckedChangeListener 设置需要实现的接口。第 21~38 行实现 CompoundButton.OnCheckedChangeListener 接口的 onCheckedChanged 方法，compoundButton.getId()可以获取到被监听对象的 id，isChecked 参数可以获取到被监听对象的开关状态。

2.10 提示信息

2.10.1 Toast 控件

Toast 控件（android.widget.Toast）是显示提示信息的 UI 控件，是 java.lang.Object 的子类。Toast 类的常量如表 2-20 所示，常用方法如表 2-21 所示。

表 2-20 Toast 类的常量

类型	常量	说明
int	LENGTH_LONG	长时间显示
int	LENGTH_SHORT	短时间显示

表 2-21 Toast 类的常用方法

类型和修饰符	方法
	Toast(Context context) 构造方法
static Toast	makeText(Context context, int resId, int duration) 创建一个基于包含 TextView 资源的 Toast 对象
static Toast	makeText(Context context, CharSequence text, int duration) 创建一个仅包含 TextView 资源的 Toast 对象
void	setDuration(int duration) 设置持续时间
void	setGravity(int gravity, int xOffset, int yOffset) 设置通知显示在屏幕上的位置
void	setMargin(float horizontalMargin, float verticalMargin) 设置视图的边距
void	setText(int resId) 设置显示的提示信息文本资源 id
void	setText(CharSequence s) 设置显示的提示信息文本
void	setView(View view) 设置作为提示信息显示的视图对象，用于自定义提示信息
void	show() 根据指定的持续时间显示提示信息
void	cancel() 如果提示信息正在显示，则取消显示

2.10.2 实例工程：不同位置显示的提示信息

本实例演示了默认位置和自定义位置显示提示信息的效果（如图 2-18 所示）。单击"显示提示信息 1"按钮会在默认位置显示提示信息，单击"显示提示信息 2"按钮会在顶部显示提示信息。

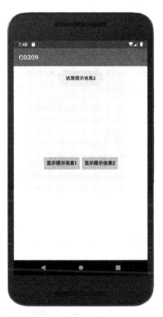

图 2-18　运行效果

1．新建工程

新建一个"Empty Activity"工程，工程名称为"C0209"。

2．主界面的布局

```
/res/layout/activity_main.xml
01  <?xml version="1.0" encoding="utf-8"?>
02  <LinearLayout xmlns:android="http://schemas.android.com/apk/res/android"
03      android:layout_width="match_parent"
04      android:layout_height="match_parent"
05      android:gravity="center">
06      <Button
07          android:id="@+id/button_show_toast1"
08          android:layout_width="wrap_content"
09          android:layout_height="wrap_content"
10          android:text="显示提示信息1" />
11      <Button
12          android:id="@+id/button_show_toast2"
13          android:layout_width="wrap_content"
14          android:layout_height="wrap_content"
15          android:text="显示提示信息2" />
16  </LinearLayout>
```

第 06～15 行添加两个 Button 控件，分别指定不同的 id 名称，用于单击后显示 Toast 提示信息。

3．主界面的 Activity

```
/java/com/vt/c0209/MainActivity.java
10  public class MainActivity extends AppCompatActivity {
11      @Override
12      protected void onCreate(Bundle savedInstanceState) {
13          super.onCreate(savedInstanceState);
14          setContentView(R.layout.activity_main);
15          Button showToastButton1 = findViewById(R.id.button_show_toast1);
16          showToastButton1.setOnClickListener(new View.OnClickListener() {
17              @Override
18              public void onClick(View v) {
19                  //直接通过静态方法调用 show()方法
20                  Toast.makeText(getApplicationContext(), "这是提示信息1", Toast.LENGTH_LONG).show();
21              }
22          });
23          Button showToastButton2 = findViewById(R.id.button_show_toast2);
24          showToastButton2.setOnClickListener(new View.OnClickListener() {
25              @Override
26              public void onClick(View v) {
27                  //使用静态方法进行赋值后再调用 show()方法
28                  Toast toast = Toast.makeText(MainActivity.this, "这是提示信息1", Toast.LENGTH_LONG);
29                  toast.setText("这是提示信息2");
30                  toast.setDuration(Toast.LENGTH_SHORT);
31                  toast.setGravity(Gravity.TOP, 0, 180);
32                  toast.show();
33              }
34          });
35      }
36  }
```

第 20 行使用 Toast 类的 makeText()静态方法直接调用 show()方法显示提示信息，由于该语句在内联方法内，不能使用 this 获取 Context 对象，需要使用 getApplicationContext()方法或 MainActivity.this 获取 Context 对象。第 28～32 行使用 Toast 类的 makeText()静态方法为 toast 对象进行赋值，再通过 3 个方法分别对文本内容、持续时间和显示位置进行修改，最后使用 show()方法显示提示信息。

 提示：Context 类

Context 类是一个抽象类，有两个子类：ContextWrapper 类和 ContextImpl 类。ContextWrapper 类是上下文功能的封装类，而 ContextImpl 类是上下文功能的实现类。ContextWrapper 类的子类包含 ContextThemeWrapper 类、Service 类和 Application 类。其中，ContextThemeWrapper 类是一个带主题的封装类，Activity 类是它的子类。

2.11 对话框

2.11.1 AlertDialog 控件

AlertDialog 控件（android.app.alertDialog）是显示对话框的 UI 控件，是 android.app.Dialog 的子类。由于 AlertDialog 类的构造方法是 protected 方法，因此需要通过 AlertDialog.Builder 类进行实例化。AlertDialog 类的常用方法如表 2-22 所示，AlertDialog.Builder 类的常用方法如表 2-23 所示。

表 2-22　AlertDialog 类的常用方法

类型和修饰符	方　　法
void	show() 显示对话框
void	cancel() 关闭对话框
void	setTitle(CharSequence title) 设置对话框标题
void	setView(View view, int viewSpacingLeft, int viewSpacingTop, int viewSpacingRight, int viewSpacingBottom) 设置对话框中显示的视图，指定该视图周围显示的间距
void	setOnShowListener(DialogInterface.OnShowListener listener) 设置在对话框显示时要调用的监听器
void	setOnCancelListener(DialogInterface.OnCancelListener listener) 设置在对话框关闭时要调用的监听器
void	setCanceledOnTouchOutside(boolean cancel) 设置触摸到对话框以外的区域是否关闭对话框

表 2-23　AlertDialog.Builder 类的常用方法

类型和修饰符	方　　法
	AlertDialog.Builder(Context context) 构造方法
AlertDialog	create() 根据对话框构建器设置的参数创建一个对话框
AlertDialog.Builder	setIcon(Drawable icon) 将 Drawable 设置为标题图标
AlertDialog.Builder	setIcon(int iconId) 根据资源 id 设置标题图标
AlertDialog.Builder	setMessage(CharSequence message) 设置要显示的消息
AlertDialog.Builder	setMessage(int messageId) 根据资源 id 设置要显示的消息
AlertDialog.Builder	setTitle(CharSequence title) 设置对话框标题
AlertDialog.Builder	setTitle(int titleId) 根据资源 id 设置标题
AlertDialog.Builder	setPositiveButton(CharSequence text, DialogInterface.OnClickListener listener) 设置肯定按钮的文字及其单击监听器

续表

类型和修饰符	方　法
AlertDialog.Builder	setNegativeButton(CharSequence text, DialogInterface.OnClickListener listener) 设置否定按钮的文字及其单击监听器
AlertDialog.Builder	setNeutralButton(CharSequence text, DialogInterface.OnClickListener listener) 设置中性按钮的文字及其单击监听器
AlertDialog.Builder	setView(View view) 设置对话框的自定义视图
AlertDialog.Builder	setSingleChoiceItems(CharSequence[] items, int checkedItem, DialogInterface.OnClickListener listener) 设置在对话框中显示的单选按钮列表和被选项，以及单击单选按钮时的监听器
AlertDialog.Builder	setMultiChoiceItems(CharSequence[] items, boolean[] checkedItems, DialogInterface.OnMultiChoiceClickListener listener) 设置在对话框中显示的复选框列表和被选项，以及单击复选框时的监听器
AlertDialog.Builder	setItems(CharSequence[] items, DialogInterface.OnClickListener listener) 设置在对话框中显示的列表，以及单击列表选项时的监听器
AlertDialog	show() 根据对话框构建器设置的参数创建一个对话框，并立即显示该对话框

 提示：单选对话框、多选对话框和列表对话框

setSingleChoiceItems()、setMultiChoiceItems()和setItems()方法可以快速构建单选对话框、多选对话框和列表对话框，但是只能使用默认样式。如果不想使用默认样式，需要使用自定义对话框。

2.11.2　实例工程：默认对话框和自定义对话框

本实例演示了默认对话框和自定义对话框的效果（如图2-19所示）。单击"默认对话框"按钮显示默认对话框，单击对话框以外的区域可以直接关闭对话框。单击"自定义对话框"按钮显示自定义对话框，然后输入框自动获取焦点后弹出虚拟键盘，此时单击对话框以外的区域，对话框不会关闭。

图2-19　运行效果

1. 新建工程

新建一个"Empty Activity"工程,工程名称为"C0210"。

2. 对话框的布局

在"/res/layout"文件夹中,新建"dialog_setting_school.xml"布局文件,用于自定义对话框的布局。

```
/res/layout/dialog_setting_school.xml
01  <?xml version="1.0" encoding="utf-8"?>
02  <LinearLayout xmlns:android="http://schemas.android.com/apk/res/android"
03      android:layout_width="match_parent"
04      android:layout_height="match_parent"
05      android:gravity="center"
06      android:orientation="vertical">
07      <EditText
08          android:id="@+id/edit_text_school"
09          android:layout_width="match_parent"
10          android:layout_height="wrap_content"
11          android:layout_marginStart="10dp"
12          android:layout_marginEnd="10dp"
13          android:gravity="center"
14          android:maxLength="16" />
15  </LinearLayout>
```

第 07~14 行添加<EditText>标签,左右外边距设置为 10dp,文字显示居中,输入文本的最大长度为 16。

3. 主界面的布局

```
/res/layout/activity_main.xml
01  <?xml version="1.0" encoding="utf-8"?>
02  <LinearLayout xmlns:android="http://schemas.android.com/apk/res/android"
03      android:layout_width="match_parent"
04      android:layout_height="match_parent"
05      android:gravity="center"
06      android:orientation="vertical">
07      <Button
08          android:id="@+id/button1"
09          android:layout_width="match_parent"
10          android:layout_height="wrap_content"
11          android:text="默认对话框" />
12      <Button
13          android:id="@+id/button2"
14          android:layout_width="match_parent"
15          android:layout_height="wrap_content"
16          android:text="自定义对话框" />
17  </LinearLayout>
```

第 07~16 行添加两个 Button 控件,指定不同的 id 名称,分别显示默认对话框和自定义对话框。

4．主界面的 Activity

```
/java/com/vt/c0210/MainActivity.java
16  public class MainActivity extends AppCompatActivity {
17      private Context mContext;
18      @Override
19      protected void onCreate(Bundle savedInstanceState) {
20          super.onCreate(savedInstanceState);
21          setContentView(R.layout.activity_main);
22          mContext = this;
23          //单击按钮显示默认对话框
24          Button button1 = findViewById(R.id.button1);
25          button1.setOnClickListener(new View.OnClickListener() {
26              @Override
27              public void onClick(View v) {
28                  showAlertDialog1();
29              }
30          });
31          //单击按钮显示自定义对话框
32          Button button2 = findViewById(R.id.button2);
33          button2.setOnClickListener(new View.OnClickListener() {
34              @Override
35              public void onClick(View v) {
36                  showAlertDialog2();
37              }
38          });
39      }
40      //显示默认对话框的方法
41      private void showAlertDialog1() {
42          //设置默认对话框的图标、标题和文本信息
43          AlertDialog.Builder adBuilder = new AlertDialog.Builder(mContext);
44          adBuilder.setIcon(R.mipmap.ic_launcher);
45          adBuilder.setTitle("默认对话框");
46          adBuilder.setMessage("这是一个默认对话框吗？");
47          //单击确认按钮事件
48          adBuilder.setPositiveButton("是", new DialogInterface.OnClickListener() {
49              @Override
50              public void onClick(DialogInterface dialog, int which) {
51                  Toast.makeText(mContext, "单击了确认按钮", Toast.LENGTH_SHORT).show();
52              }
53          });
54          //单击取消按钮事件
55          adBuilder.setNegativeButton("否", new DialogInterface.OnClickListener() {
56              @Override
57              public void onClick(DialogInterface dialog, int which) {
58                  Toast.makeText(mContext, "单击了取消按钮", Toast.LENGTH_SHORT).show();
59              }
60          });
61          //单击中性按钮事件
62          adBuilder.setNeutralButton("不确定", new DialogInterface.OnClickListener() {
```

```
63              @Override
64              public void onClick(DialogInterface dialog, int which) {
65                  Toast.makeText(mContext, "单击了中性按钮", Toast.LENGTH_SHORT).show();
66              }
67          });
68          adBuilder.show();
69      }
70      //显示自定义对话框的方法
71      private void showAlertDialog2() {
72          //通过 LayoutInflater 加载 XML 布局文件作为一个 View 对象
73          View view = LayoutInflater.from(mContext).inflate(R.layout.dialog_setting_school, null);
74          //设置自定义对话框的标题及布局文件
75          AlertDialog.Builder adBuilder = new AlertDialog.Builder(mContext);
76          adBuilder.setTitle("自定义对话框:请输入院校的全称");
77          adBuilder.setView(view);
78          //获取自定义布局中的 EditText 控件
79          final EditText schoolEditText = view.findViewById(R.id.edit_text_school);
80          //单击确定按钮事件:对 EditText 输入的内容进行判断
81          adBuilder.setPositiveButton("确定", new DialogInterface.OnClickListener() {
82              @Override
83              public void onClick(DialogInterface dialog, int which) {
84                  String school = schoolEditText.getText().toString().trim();
85                  if (!school.equals("")) {
86                      if (school.endsWith("大学") || school.endsWith("学院") || school.endsWith("学校")) {
87                          Toast.makeText(mContext, school, Toast.LENGTH_LONG).show();
88                      } else {
89                          Toast.makeText(mContext, "请使用院校标准名称! ", Toast.LENGTH_LONG).show();
90                      }
91                  } else {
92                      Toast.makeText(mContext, "院校未设置", Toast.LENGTH_LONG).show();
93                  }
94              }
95          });
96          adBuilder.setNegativeButton("取消", null);  //未设置单击监听器
97          //设置对话框显示的位置并显示对话框
98          AlertDialog alertDialog = adBuilder.create();
99          alertDialog.setOnShowListener(new DialogInterface.OnShowListener() {
100             public void onShow(DialogInterface dialog) {
101                 //100毫秒后 EditText 获取焦点并弹出虚拟键盘
102                 new Handler().postDelayed(new Runnable() {
103                     @Override
104                     public void run() {
105                         schoolEditText.requestFocus();//EditText 获取焦点
106                         showInputMethod();//弹出虚拟键盘
107                     }
108                 }, 100);
109             }
110         });
111         alertDialog.setView(view, 0, 50, 0, 50);
112         alertDialog.setCanceledOnTouchOutside(false);//单击对话框以外的区域不消失
```

```
113            alertDialog.show();
114        }
115    //弹出虚拟键盘的方法
116    private void showInputMethod() {
117        InputMethodManager inputMethodManager = (InputMethodManager) mContext.getSystemService
    (Context.INPUT_METHOD_SERVICE);
118        inputMethodManager.toggleSoftInput(0, InputMethodManager.HIDE_NOT_ALWAYS);
119    }
120 }
```

第 41~69 行添加 showAlertDialog1()方法显示默认对话框，使用 AlertDialog.Builder 类构建对话框，设置 3 个内置的按钮及其单击监听事件，使用 show()方法根据设置的参数创建对话框，并立即显示该对话框，并没有直接使用 AlertDialog 类。第 71~114 行添加 showAlertDialog2()方法显示自定义对话框，通过 AlertDialog.Builder 类的 setView()方法设置对话框的自定义布局，再使用 AlertDialog.Builder 类的 create()方法初始化 AlertDialog 类实例，并使用 setOnShowListener()方法在对话框显示后使 EditText 控件获取弹出虚拟键盘的 showInputMethod()方法，然后使用 AlertDialog 类的 setView()方法重新设置自定义布局，与 AlertDialog.Builder 类的 setView()方法相比可以多设置自定义布局周围的间距，最后使用 AlertDialog 类的 show()方法显示对话框。第 116~119 行添加 showInputMethod()方法，实现弹出虚拟键盘。

2.12 日期选择器

2.12.1 DatePicker 控件

DatePicker 控件（android.widget.DatePicker）是选择日期的 UI 控件，是 android.widget.FrameLayout 的子类。DatePicker 控件的常用 XML 标签属性如表 2-24 所示 DatePicker 类的常用方法如表 2-25 所示。

表 2-24 DatePicker 控件的常用 XML 标签属性

属性	说明
android:calendarTextColor	设置日历列表的文本颜色
android:calendarViewShown	设置是否显示日历视图
android:datePickerMode	设置外观样式，可选值为 spinner 和 calendar。calendar 是默认值，需要较大的显示面积，因此一般使用 spinner 选择日期
android:maxDate	设置最大日期，格式为 mm/dd/yyyy
android:minDate	设置最小日期，格式为 mm/dd/yyy
android:spinnersShown	设置是否显示 spinner
android:startYear	设置起始年份
android:endYear	设置结束年份

表 2-25　DatePicker 类的常用方法

类型和修饰符	方　　法
void	init(int year, int month, int day, DatePicker.OnDateChangedListener onDateChangedListener) 初始化默认日期及日期改变的监听器
void	setOnDateChangedListener(DatePicker.OnDateChangedListener onDateChangedListener) 设置改变日期的监听器
void	updateDate(int year, int month, int dayOfMonth) 更新日期，月份从 0 开始

2.12.2　实例工程：设置日期的日期选择器

本实例演示了 spinner 外观样式的日期选择器（如图 2-20 所示）。滑动 DatePicker 控件可以修改年月日。单击"重置"按钮，会将 DatePicker 控件的日期设置为"2020 年 1 月 1 日"。单击"确定"按钮，会获取 DatePicker 控件设置的日期。

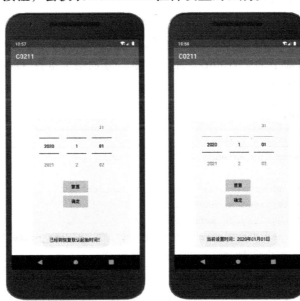

图 2-20　运行效果

1．新建工程

新建一个"Empty Activity"工程，工程名称为"C0211"。

2．主界面的布局

```
/res/layout/activity_main.xml
01   <?xml version="1.0" encoding="utf-8"?>
02   <LinearLayout xmlns:android="http://schemas.android.com/apk/res/android"
03       android:layout_width="match_parent"
04       android:layout_height="match_parent"
05       android:gravity="center"
06       android:orientation="vertical">
07       <DatePicker
```

```
08              android:id="@+id/date_picker"
09              android:layout_width="match_parent"
10              android:layout_height="wrap_content"
11              android:calendarViewShown="false"
12              android:datePickerMode="spinner"
13              android:startYear="2020" />
14          <Button
15              android:id="@+id/button1"
16              android:layout_width="wrap_content"
17              android:layout_height="wrap_content"
18              android:text="重置" />
19          <Button
20              android:id="@+id/button2"
21              android:layout_width="wrap_content"
22              android:layout_height="wrap_content"
23              android:text="确定" />
24      </LinearLayout>
```

第 07～13 行添加了<DatePicker>控件标签，设置为 spinner 模式、隐藏日历视图和起始年。第 14～23 行添加了两个<Button>控件标签，分别用于重置 DatePicker 控件和显示 DatePicker 控件选择的日期。

3. 主界面的 Activity

```
/java/com/vt/c0211/MainActivity.java
12  public class MainActivity extends AppCompatActivity {
13      @Override
14      protected void onCreate(Bundle savedInstanceState) {
15          super.onCreate(savedInstanceState);
16          setContentView(R.layout.activity_main);
17          final DatePicker datePicker = findViewById(R.id.date_picker);
18          datePicker.updateDate(2020, 0, 1);
19          datePicker.setOnDateChangedListener(new DatePicker.OnDateChangedListener() {
20              @Override
21              public void onDateChanged(DatePicker view, int year, int monthOfYear, int dayOfMonth) {
22                  Calendar calendar = Calendar.getInstance();
23                  calendar.set(year, monthOfYear, dayOfMonth);
24                  SimpleDateFormat simpleDateFormat = new SimpleDateFormat("yyyy年MM月dd日");
25                  Toast.makeText(getApplicationContext(), "当前选择时间: " + simpleDateFormat.format(calendar.getTime()), Toast.LENGTH_SHORT).show();
26              }
27          });
28          Button button1 = findViewById(R.id.button1);
29          button1.setOnClickListener(new View.OnClickListener() {
30              @Override
31              public void onClick(View v) {
32                  datePicker.updateDate(2020, 0, 1);
33                  Toast.makeText(getApplicationContext(), "已经到恢复默认起始时间！", Toast.LENGTH_LONG).show();
34              }
35          });
```

```
36          Button button2 = findViewById(R.id.button2);
37          button2.setOnClickListener(new View.OnClickListener() {
38              @Override
39              public void onClick(View v) {
40                  Calendar calendar = Calendar.getInstance();
41                  calendar.set(datePicker.getYear(), datePicker.getMonth(), datePicker.getDayOfMonth());
42                  SimpleDateFormat simpleDateFormat = new SimpleDateFormat("yyyy年MM月dd日");
43                  Toast.makeText(getApplicationContext(), "当前设置时间: " + simpleDateFormat.
    format(calendar.getTime()), Toast.LENGTH_SHORT).show();
44              }
45          });
46      }
47  }
```

第 18 行设置 DatePicker 控件的日期，未设置时间时会以本地时间作为默认时间。第 19～27 行设置修改日期监听器，通过 Calendar 类合成时间，并通过 SimpleDateFormat 类进行时间格式化，然后使用提示信息将其显示出来。第 32 行重新设置 DatePicker 控件的日期。第 40～43 行获取 DatePicker 控件的日期，然后使用提示信息将其显示出来。

 提示：DatePickerDialog 类

DatePickerDialog 类是 android.app.AlertDialog 的子类，相当于在 AlertDialog 控件上加载了 DatePicker 控件，与自定义对话框相比更加快捷。

2.13 时间选择器

2.13.1 TimePicker 控件

TimePicker 控件（android.widget.TimePicker）是选择时间的 UI 控件，是 android.widget.FrameLayout 的子类。TimePicker 控件的常用 XML 标签属性如表 2-26 所示。TimePicker 类的常用方法如表 2-27 所示。

表 2-26 TimePicker 控件的常用 XML 标签属性

属性	说明
android:timePickerMode	设置外观样式，可选值为 spinner 和 clock。clock 是默认值，以模拟表盘的方式选取时间，也可以直接输入数值

表 2-27 TimePicker 类的常用方法

类型和修饰符	方法
int	getMinute() 获取选择的分钟
int	getHour() 获取选择的小时
boolean	is24HourView() 判断是否以 24 小时制显示时间

续表

类型和修饰符	方　　法
void	setMinute(int minute) 设置当前选择的分钟
void	setHour(int hour) 使用 24 小时制设置当前选择的小时
void	setIs24HourView(Boolean is24HourView) 设置是否以 24 小时制显示时间
void	setOnTimeChangedListener(TimePicker.OnTimeChangedListener onTimeChangedListener) 设置改变时间的监听器

2.13.2　实例工程：设置时间的时间选择器

本实例演示了 12 小时制和 24 小时制的时间选择器（如图 2-21 所示）。第一个 DatePicker 控件使用 24 小时制；第二个 DatePicker 控件使用 12 小时制，增加了"上午"和"下午"的选择。

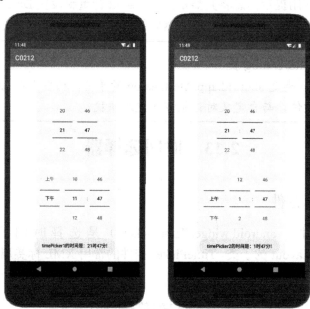

图 2-21　运行效果

1．新建工程

新建一个"Empty Activity"工程，工程名称为"C0212"。

2．主界面的布局

```
/res/layout/activity_main.xml
01  <?xml version="1.0" encoding="utf-8"?>
02  <LinearLayout xmlns:android="http://schemas.android.com/apk/res/android"
03      android:layout_width="match_parent"
04      android:layout_height="match_parent"
05      android:gravity="center"
```

```
06        android:orientation="vertical">
07        <TimePicker
08            android:id="@+id/time_picker1"
09            android:layout_width="match_parent"
10            android:layout_height="wrap_content"
11            android:timePickerMode="spinner" />
12        <TimePicker
13            android:id="@+id/time_picker2"
14            android:layout_width="match_parent"
15            android:layout_height="wrap_content"
16            android:timePickerMode="spinner" />
17    </LinearLayout>
```

第 07～16 行添加两个 TimePicker 控件，除 id 外其他属性的设置都相同，用于对比 12 小时制和 24 小时制的区别。由于 24 小时制显示时间没有标签属性，因此需要通过类方法进行设置。

3．主界面的 Activity

```
/java/com/vt/c0212/MainActivity.java
08  public class MainActivity extends AppCompatActivity {
09      @Override
10      protected void onCreate(Bundle savedInstanceState) {
11          super.onCreate(savedInstanceState);
12          setContentView(R.layout.activity_main);
13          //24 小时制显示时间
14          TimePicker timePicker1 = findViewById(R.id.time_picker1);
15          timePicker1.setIs24HourView(true);
16          timePicker1.setOnTimeChangedListener(new TimePicker.OnTimeChangedListener() {
17              @Override
18              public void onTimeChanged(TimePicker view, int hourOfDay, int minute) {
19                  Toast.makeText(MainActivity.this, "timePicker1 的时间是: " + hourOfDay + "时" + minute + "分!", Toast.LENGTH_SHORT).show();
20              }
21          });
22          //12 小时制显示时间
23          TimePicker timePicker2 = findViewById(R.id.time_picker2);
24          timePicker2.setOnTimeChangedListener(new TimePicker.OnTimeChangedListener() {
25              @Override
26              public void onTimeChanged(TimePicker view, int hourOfDay, int minute) {
27                  Toast.makeText(MainActivity.this, "timePicker2 的时间是: " + hourOfDay + "时" + minute + "分!", Toast.LENGTH_SHORT).show();
28              }
29          });
30      }
31  }
```

第 14～21 行将 time_Picker1 设置为 24 小时制显示，并设置改变时间的监听器用于显示改变后的时间。第 23～29 行 time_Picker2 设置为 12 小时制显示，并设置改变时间的监听器用于显示改变后的时间。

提示：TimePickerDialog 类

TimePickerDialog 类是 android.app.AlertDialog 的子类，相当于在 AlertDialog 控件上加载了 TimePicker 控件，与自定义对话框相比更加快捷。

2.14 滚动条视图

2.14.1 ScrollView 控件

ScrollView 控件（android.widget.ScrollView）是可滚动的显示其他控件的 UI 控件，是 android.widget.FrameLayout 的子类。ScrollView 控件只能包含一个子控件，否则会报错，但是在子控件中可以再添加控件。ScrollView 控件的常用 XML 标签属性如表 2-28 所示，ScrollView 类的常用方法如表 2-29 所示。

表 2-28 ScrollView 控件的常用 XML 标签属性

属性	说明
android:fillViewport	设置是否拉伸其内容以填充视图。在嵌套的子控件高度达不到屏幕高度时，即使 ScrollView 高度设置了 match_parent，也无法充满整个屏幕，需要改属性值为 true，使 ScrollView 填充整个页面
android:scrollbars	设置在滚动时是否显示滚动条。可选值为 none、horizontal 和 vertical
android:scrollbarThumbHorizontal	设置水平方向的滑块图像
android:scrollbarThumbVertical	设置垂直方向的滑块图像
android:descendantFocusability	设置当获取焦点时和其子控件之间的关系。属性值有以下 3 种。 ● beforeDescendants：优先其子控件而获取到焦点 ● afterDescendants：其子控件无须获取焦点时才获取焦点 ● blocksDescendants：覆盖子控件而直接获得焦点

表 2-29 ScrollView 类的常用方法

类型和修饰符	方法
void	addView(View child) 添加子 View 控件
final int	getScrollX() 获取水平滚动位置
final int	getScrollY() 获取垂直滚动位置
void	scrollTo(int x, int y) 立即滚动到指定的坐标位置，不能超出子控件的范围
void	scrollBy(int x, int y) 立即滚动到指定的相对坐标位置，不能超出子控件的范围
void	smoothScrollTo(int x, int y) 平滑滚动到指定的坐标位置，不能超出子控件的范围
void	smoothScrollBy(int dx, int dy) 平滑滚动到指定的相对坐标位置，不能超出子控件的范围
void	scrollToDescendant(View child) 滚动显示出指定的子控件。如果该子控件已经显示出来，则无效果

续表

类型和修饰符	方 法
boolean	post(Runnable action) 将 Runnable 添加到消息队列中,在视图绘制完成后才执行
void	computeScroll() 由父控件调用,请求子视图根据偏移值 mScrollX 和 mScrollY 重新绘制

2.14.2　实例工程:滚动显示视图

本实例演示了子控件超出滚动条显示范围时通过滚动的方式进行显示(如图 2-22 所示)。ScrollView 控件先向下滑动到 1000 像素,然后单击其下方的 4 个按钮实现不同位置的滑动。

图 2-22　运行效果

1.新建工程

新建一个"Empty Activity"工程,工程名称为"C0213"。

2.主界面的布局

```
/res/layout/activity_main.xml
01  <?xml version="1.0" encoding="utf-8"?>
02  <LinearLayout xmlns:android="http://schemas.android.com/apk/res/android"
03      android:layout_width="match_parent"
04      android:layout_height="match_parent"
05      android:orientation="vertical">
06      <ScrollView
07          android:id="@+id/scroll_view"
08          android:layout_width="match_parent"
09          android:layout_height="520dp">
10          <LinearLayout
11              android:id="@+id/linear_layout"
12              android:layout_width="match_parent"
```

```xml
13          android:layout_height="wrap_content"
14          android:orientation="vertical" />
15    </ScrollView>
16    <LinearLayout
17        android:layout_width="wrap_content"
18        android:layout_height="wrap_content"
19        android:layout_gravity="center"
20        android:orientation="horizontal">
21        <Button
22            android:id="@+id/button1"
23            android:layout_width="wrap_content"
24            android:layout_height="wrap_content"
25            android:text="滑动到顶部" />
26        <Button
27            android:id="@+id/button2"
28            android:layout_width="wrap_content"
29            android:layout_height="wrap_content"
30            android:text="滑动到底部" />
31        <Button
32            android:id="@+id/button3"
33            android:layout_width="wrap_content"
34            android:layout_height="wrap_content"
35            android:text="滑动到指定控件" />
36        <Button
37            android:id="@+id/button4"
38            android:layout_width="wrap_content"
39            android:layout_height="wrap_content"
40            android:text="随机滑动" />
41    </LinearLayout>
42 </LinearLayout>
```

第 06～15 行添加一个 ScrollView 控件，其 LinearLayout 子控件用于添加其他控件。第 16～41 行在一个 LinearLayout 控件中添加 4 个 Button 控件，用于实现不同的跳转滑动效果。

3．主界面的 Activity

```java
/java/com/vt/c0213/MainActivity.java
11  public class MainActivity extends AppCompatActivity {
12      @Override
13      protected void onCreate(Bundle savedInstanceState) {
14          super.onCreate(savedInstanceState);
15          setContentView(R.layout.activity_main);
16          //scrollView 在视图上加载后执行
17          final ScrollView scrollView = findViewById(R.id.scroll_view);
18          scrollView.post(new Runnable() {
19              @Override
20              public void run() {
21                  scrollView.smoothScrollTo(0,1000);
22              }
23          });
24          //scrollView 上添加 50 个 TextView 控件
```

```
25      final TextView[] textView = new TextView[50];
26      LinearLayout linearLayout = findViewById(R.id.linear_layout);
27      for (int i = 0; i < 50; i++) {
28          //实例化TextView数组的元素
29          textView[i] = new TextView(this);
30          textView[i].setText(String.format("第%d个TextView控件", 1 + i));
31          textView[i].setTextSize(30);
32          linearLayout.addView(textView[i]);
33      }
34      //scrollView滑动到顶部
35      Button button1 = findViewById(R.id.button1);
36      button1.setOnClickListener(new View.OnClickListener() {
37          @Override
38          public void onClick(View v) {
39              scrollView.fullScroll(ScrollView.FOCUS_UP);
40          }
41      });
42      //scrollView滑动到底部
43      Button button2 = findViewById(R.id.button2);
44      button2.setOnClickListener(new View.OnClickListener() {
45          @Override
46          public void onClick(View v) {
47              scrollView.fullScroll(ScrollView.FOCUS_DOWN);
48          }
49      });
50      //scrollView滑动到指定控件
51      Button button3 = findViewById(R.id.button3);
52      button3.setOnClickListener(new View.OnClickListener() {
53          @Override
54          public void onClick(View v) {
55              scrollView.scrollToDescendant(textView[20]);
56          }
57      });
58      //scrollView滑动到随机位置
59      Button button4 = findViewById(R.id.button4);
60      button4.setOnClickListener(new View.OnClickListener() {
61          @Override
62          public void onClick(View v) {
63              scrollView.smoothScrollTo(0,(int)(Math.random()*5000));
64          }
65      });
66  }
67 }
```

第 18~23 行在 ScrollView 控件加载后执行 Runnable 类的 run()方法。第 29~32 行 TextView 数组的元素进行实例化后，为其设置属性再添加到 LinearLayout 控件中。第 39 行 ScrollView.FOCUS_UP 是指移动焦点到顶部。第 47 行 ScrollView.FOCUS_DWON 是指移动焦点到底部。第 55 行 textView[20]是指 textView[]数组中的第 21 个 TextView 元素。第 63 行 Math.random()*5000 生成一个 5000 以内的随机数并使用(int)强制转换成 int 型数值。

2.15 通　　知

2.15.1 Notification 控件

Notification 控件（android.app.Notification）是显示通知的 UI 控件，是 java.lang.Object 的子类。当 App 向系统发出通知时，先以图标的形式显示在通知栏中。当用户下拉通知栏后，可以查看通知的详细信息（如图 2-23 所示）。

图 2-23　通知详细信息的组成

Notification 类没有提供创建通知的方法，需要使用 Notification.Builder 类（如表 2-30 所示）构建通知，NotificationManager 类管理通知（如表 2-31 所示），NotificationChannel 类创建通知的通道（如表 2-32 所示）。

表 2-30　Notification.Builder 类的常用方法

类型和修饰符	方　　法
Notification.Builder	Notification.Builder(Context context, String channelId) 构造方法
Notification.Builder	setAutoCancel(boolean autoCancel) 设置用户触摸时状态栏通知是否自动取消
Notification.Builder	setBadgeIconType(int icon) 设置状态栏通知的 badge 图标类型。可选类型包括 Notification.BADGE_ICON_NONE、Notification.BADGE_ICON_SMALL、Notification.BADGE_ICON_LARGE
Notification.Builder	setChannelId(String channelId) 设置通知发送的通道 id
Notification.Builder	setContentIntent(PendingIntent intent) 设置单击通知时要发送的 PendingIntent
Notification.Builder	setContentText(CharSequence text) 设置通知的第二行文本
Notification.Builder	setContentTitle(CharSequence title) 设置通知的第一行文本
Notification.Builder	setCustomBigContentView(RemoteViews contentView) 设置大尺寸的自定义远程视图
Notification.Builder	setCustomContentView (RemoteViews contentView) 设置自定义远程视图
Notification.Builder	setCustomHeadsUpContentView (RemoteViews contentView) 设置自定义弹出式远程视图，自动消失后不在状态栏显示通知
Notification.Builder	setExtras(Bundle extras) 设置附加数据。每次调用 build() 时，Bundle 都会复制到通知中

类型和修饰符	方 法
Notification.Builder	setFullScreenIntent(PendingIntent intent, boolean highPriority) 当 highPriority 为 true 时，在其他 App 全屏状态时直接 PendingIntent；当 highPriority 为 false 时，在状态栏显示通知
Notification.Builder	setLargeIcon(Bitmap b) 设置通知的大图标
Notification.Builder	setNumber(int number) 设置通知表示的项目数，当支持 badge 时作为 badge 数量显示
Notification.Builder	setOngoing(boolean ongoing) 设置是否为正在进行的通知。用户无法通过界面操作取消正在进行的通知，通常用于播放音乐、文件下载、同步操作、活动网络连接等
Notification.Builder	setOnlyAlertOnce(boolean onlyAlertOnce) 设置通知是否只提示一次
Notification.Builder	setSmallIcon(int icon) 设置通知的小图标
Notification.Builder	setSubText(CharSequence text) 设置通知的附加信息
Notification.Builder	setTicker(CharSequence tickerText) 设置辅助功能的文本
Notification.Builder	setTimeoutAfter (long durationMs) 设置通知自动消失的时间
Notification.Builder	setVisibility (int visibility) 设置锁屏状态时通知的可见属性，可选值如下： ● Notification.VISIBILITY_PUBLIC，锁屏时显示通知 ● Notification.VISIBILITY_PRIVATE，锁屏时显示通知，但在安全锁屏时隐藏敏感或私有信息 ● Notification.VISIBILITY_SECRET，安全锁屏时不显示通知

表 2-31　NotificationManager 类的常用方法

类型和修饰符	方 法
void	createNotificationChannel(NotificationChannel channel) 创建通知通道
void	cancel(int id) 取消显示指定 id 的通知
void	cancel(String tag, int id) 取消显示指定 tag 和 id 的通知
void	cancelAll() 取消所有显示的通知
void	deleteNotificationChannel(String channelId) 删除指定 id 的通知通道
void	notify(int id, Notification notification) 发送通知，为该通知指定 id
void	notify(String tag, int id, Notification notification) 发送通知，为该通知指定 tag 和 id

表 2-32　NotificationChannel 类的常用方法

类型和修饰符	方 法
	NotificationChannel(String id, CharSequence name, int importance) 构造方法

续表

类型和修饰符	方　　法
void	enableLights(boolean lights) 是否允许通知灯
void	enableVibration(boolean vibration) 是否允许通知震动
void	setAllowBubbles(boolean allowBubbles) 是否允许 Bubble
void	setImportance(int importance) 设置通知级别。可选值如下： ● NotificationManager.IMPORTANCE_UNSPECIFIED，未表示重要性的通知 ● NotificationManager.IMPORTANCE_NONE，不在状态栏中显示 ● NotificationManager.IMPORTANCE_MIN，最小化的通知，折叠在状态栏中 ● NotificationManager.IMPORTANCE_LOW，低级别的通知，显示在状态栏的折叠通知中，并可能显示在状态栏中，没有提示音 ● NotificationManager.IMPORTANCE_DEFAULT，普通重要通知，有提示音 ● NotificationManager.IMPORTANCE_HIGH，高级别重要通知，弹出式显示，可以在全屏下显示
void	setShowBadge(boolean showBadge) 是否在 App 启动图标上显示 badge
void	setSound(Uri sound, AudioAttributes audioAttributes) 设置通知的提示音
void	setVibrationPattern(long[] vibrationPattern) 设置通知的震动间隔
boolean	shouldShowLights() 获取是否触发通知灯
boolean	shouldVibrate() 获取是否触发震动

2.15.2　实例工程：弹出式状态栏通知和自定义视图状态栏通知

本实例演示了弹出式状态栏通知和自定义视图状态栏通知（如图 2-24 所示），单击"关闭状态栏通知"按钮可以关闭通知。

图 2-24　运行效果

1. 打开基础工程

打开"基础工程"文件夹中的"C0214"工程，该工程已经包含图像资源、MainActivity 及布局的文件。

2. 主界面的 Activity

```
/java/com/vt/c0214/MainActivity.java
```

```java
17  public class MainActivity extends AppCompatActivity implements View.OnClickListener{
18      private final String NOTIFY_TAG = "通知";
19      private final String CHANNL_ID1 = "2401";
20      private final String CHANNL_NAME1 = "通知";
21      private final String CHANNL_ID2 = "2402";
22      private final String CHANNL_NAME2 = "对话";
23      private Context mContext;
24      private NotificationManager mNotificationManager;
25      private Notification.Builder mNotificationBuilder;
26      private Notification mNotification;
27      @Override
28      protected void onCreate(Bundle savedInstanceState) {
29          super.onCreate(savedInstanceState);
30          setContentView(R.layout.activity_main);
31          mContext = MainActivity.this;
32          mNotificationManager = (NotificationManager) getSystemService(NOTIFICATION_SERVICE);
33          findViewById(R.id.btn_show_notification).setOnClickListener(this);
34          findViewById(R.id.btn_show_custom_notification).setOnClickListener(this);
35          findViewById(R.id.btn_close_notification).setOnClickListener(this);
36      }
37      //单击事件监听器
38      @Override
39      public void onClick(View v) {
40          switch (v.getId()) {
41              case R.id.btn_show_notification:
42                  //兼容 Android 8.0(API level 26)以后的版本
43                  if (Build.VERSION.SDK_INT >= Build.VERSION_CODES.O) {
44                      NotificationChannel notificationChannel = new NotificationChannel
45  (CHANNL_ID1, CHANNL_NAME1, NotificationManager.IMPORTANCE_HIGH);
                        notificationChannel.setShowBadge(true);
46                      notificationChannel.enableVibration(true);
47                      notificationChannel.enableLights(true);
48                      notificationChannel.setSound(RingtoneManager.getDefaultUri
49  (RingtoneManager.TYPE_NOTIFICATION), Notification.AUDIO_ATTRIBUTES_DEFAULT);
                        notificationChannel.setVibrationPattern(new long[]{100, 200, 300,
50  400, 500, 400, 300, 200, 400});
                        mNotificationManager.createNotificationChannel(notificationChannel);
51                      mNotificationBuilder = new Notification.Builder(mContext, CHANNL_ID1);
52                  } else {
53                      mNotificationBuilder = new Notification.Builder(mContext);
54                  }
55                  //设置通知参数
56                  mNotificationBuilder.setContentTitle("通知标题: 会议通知")
```

```
57                        .setContentText("通知内容：周五下午1：30在会议室召开全体会议。")
58                        .setTicker("收到一条会议信息")
59                        .setSmallIcon(R.mipmap.notification_small)
60                        .setLargeIcon(BitmapFactory.decodeResource(getResources(),
    R.mipmap.notification_icon))
61                        .setBadgeIconType(Notification.BADGE_ICON_SMALL)
62                        .setNumber(11)
63                        .setSubText("重要提醒")
64                        .setAutoCancel(true);
65                 mNotification = mNotificationBuilder.build();
66                 //发出通知
67                 mNotificationManager.notify(NOTIFY_TAG, Integer.valueOf(CHANNL_ID1), mNotification);
68                 break;
69             case R.id.btn_show_custom_notification:
70                 //兼容 Android 8.0(API level 26)以后的版本
71                 if (Build.VERSION.SDK_INT >= Build.VERSION_CODES.O) {
72                     NotificationChannel notificationChannel = new NotificationChannel
    (CHANNL_ID2, CHANNL_NAME2, NotificationManager.IMPORTANCE_HIGH);
73                     notificationChannel.setShowBadge(true);
74                     notificationChannel.setVibrationPattern(new long[]{100, 200, 300,
75                     400, 500, 400, 300, 200, 400});
                        mNotificationManager.createNotificationChannel(notificationChannel);
76                     mNotificationBuilder = new Notification.Builder(mContext, CHANNL_ID2);
77                 } else {
78                     mNotificationBuilder = new Notification.Builder(mContext);
79                 }
80                 //设置自定义视图
81                 PendingIntent pendingIntent = PendingIntent.getActivity(mContext, 1, new
    Intent(mContext, MainActivity.class), PendingIntent.FLAG_UPDATE_CURRENT);
82                 RemoteViews remoteView = new RemoteViews(getPackageName(), R.layout.
    notification_custom);
83                 remoteView.setImageViewResource(R.id.user_face, R.mipmap.img_user_face);
84                 remoteView.setOnClickPendingIntent(R.id.btn_reply, pendingIntent);
85                 //设置通知参数
86                 mNotificationBuilder.setContentTitle("通知标题：会议通知")
87                        .setContentText("通知内容：周五下午1：30在会议室召开全体会议。")
88                        .setTicker("收到一条会议信息")
89                        .setSmallIcon(R.mipmap.notification_small)
90                        .setLargeIcon(BitmapFactory.decodeResource(getResources(),
    R.mipmap.notification_icon))
91                        .setNumber(10)
92                        .setVisibility(Notification.VISIBILITY_PUBLIC)
93                        .setCustomContentView(remoteView);
94                 mNotification = mNotificationBuilder.build();
95                 //发出通知
96                 mNotificationManager.notify(NOTIFY_TAG, Integer.valueOf(CHANNL_ID2), mNotification);
97                 break;
98             case R.id.btn_close_notification:
99                 //除了可以根据 ID 来取消 Notification,还可以调用 cancelAll()关闭该应用的所有通知
100                mNotificationManager.cancel(NOTIFY_TAG, Integer.valueOf(CHANNL_ID1));
```

```
101              mNotificationManager.cancel(NOTIFY_TAG, Integer.valueOf(CHANNL_ID2));
102              mNotificationManager.deleteNotificationChannel(CHANNL_ID1);
103              mNotificationManager.deleteNotificationChannel(CHANNL_ID2);
104              break;
105         }
106     }
107 }
```

第 18～22 行定义通知需要使用的常量。第 32 行 NotificationManager 类没有构造方法，需要使用（NotificationManager）getSystemService(NOTIFICATION_SERVICE) 初始化 NotificationManager 实例。第 43 行 Build.VERSION.SDK_INT 表示生成 App 使用的 SDK 版本，Build.VERSION_CODES.O 表示使用 Android 版本（API level 26）的 SDK。第 44～49 行创建通知通道的实例，并设置相关属性。第 56～65 行设置通知所需的相关属性，然后使用 build()方法构建通知。第 81 行创建 PendingIntent 实例作为自定义视图中"回复"按钮的单击时触发的对象。第 82～84 行创建 RemoteViews 实例，用于设置自定义视图，然后在第 93 行设置该实例作为通知的自定义视图。

> 提示：RemoteViews
> RemoteViews 只需要包名和待加载的资源文件 id，并不支持所有类型的 View，也不支持自定义的 View，支持的类型如下。
> - Layout: FrameLayout、LinearLayout、RelativeLayout 和 GridLayout。
> - View: Button、ImageView、ImageButton、ProgressBar、TextView、ListView、GridView、StackView、ViewStub、AdapterViewFlipper、ViewFlipper、AnalogClock 和 Chronometer。

2.16 习　　题

1. 实现全屏显示图像的界面效果。
2. 实现多选题的界面效果。
3. 实现选择出生日期的界面效果。

第 3 章 UI 布局控件

虽然 UI 控件可以不依靠 UI 布局控件单独使用，但是在实际使用中，几乎很少只使用一个 UI 控件。当多个 UI 控件同时使用时，需要通过 UI 布局控件组织控件的排列方式和位置。相同的界面效果可以使用不同的 UI 布局控件实现，因此需要熟悉各种 UI 布局控件的特点，选择最适合的 UI 布局控件。

3.1 线 性 布 局

3.1.1 LinearLayout 控件

LinearLayout 控件（android.widget.LinearLayout）是进行水平或垂直排列布局的 UI 控件，是 android.view.ViewGroup 的子类。LinearLayout 控件的常用 XML 标签属性如表 3-1 所示，LinearLayout 子控件的 XML 标签属性如表 3-2 所示。

表 3-1　LinearLayout 控件的常用 XML 标签属性

属　性	说　明
android:divider	设置分割线的图像资源
android:gravity	设置控件内部的对齐方式
android:measureWithLargestChild	设置是否以最大尺寸的子控件作为所有子控件的尺寸
android:orientation	设置线性布局的排列方式，可选值包括 horizontal 和 vertical
android:weightSum	设置权重值的最大和

表 3-2　LinearLayout 子控件的 XML 标签属性

属　性	说　明
android:layout_gravity	设置在父控件中的定位方式
android:layout_weight	设置子控件在父控件中所占的权重；子控件权重除以其父控件中所有子控件权重之和的值是分配剩余控件空间的比例

 提示：gravity 和 layout_gravity

android:gravity 是针对控件中的元素而言的，用来控制元素在该控件中的显示位置。当 Button 控件设置 android:gravity="left" 时，Button 控件上的文字将位于内部左侧。

android:layout_gravity 是针对控件本身而言的，用来控制该控件在包含该控件的父控件中的位置。当 Button 控件设置 android:layout_gravity="left" 时，Button 控件将位于父控件中的左侧。

提示：RTL 布局

从 Android 4.2 开始，Android SDK 支持一种从右到左（right-to-left，RTL）的 UI 布局方式，以右边上角作为原点，从右到左、从上到下进行排列，经常使用在阿拉伯语、希伯来语等环境中。如果要使用 RTL 布局，需要在 AndroidManifest.xml 文件中将<application>标签的 android:supportsRtl 属性值设为 true，将控件标签的 android:layoutDirection 属性值设为 rtl。

3.1.2 实例工程：动态视图的线性布局

本实例演示了水平和垂直两种线性布局及嵌套布局的效果（如图 3-1 所示）。整体使用垂直线性布局，局部使用多重嵌套的线性布局。

1. 新建工程并导入素材

新建一个"Empty Activity"工程，工程名称为"C0301"。选择"/res/drawable"资源文件夹，将素材图像（素材文件夹路径为"/素材/C0301"）粘贴或拖曳到该文件夹中。

2. 主界面的布局

图 3-1 运行效果

```
/res/layout/activity_main.xml
01  <?xml version="1.0" encoding="utf-8"?>
02  <!--垂直线性布局-->
03  <LinearLayout xmlns:android="http://schemas.android.com/apk/res/android"
04      android:layout_width="match_parent"
05      android:layout_height="wrap_content"
06      android:orientation="vertical">
07      <!--水平线性布局-->
08      <LinearLayout
09          android:layout_width="match_parent"
10          android:layout_height="50dp"
11          android:layout_margin="5dp"
12          android:orientation="horizontal">
13          <ImageView
14              android:layout_width="50dp"
15              android:layout_height="50dp"
16              android:scaleType="centerCrop"
17              android:src="@mipmap/img_face" />
18          <!--垂直线性布局-->
19          <LinearLayout
20              android:layout_width="0dp"
21              android:layout_height="match_parent"
22              android:layout_marginStart="5dp"
23              android:layout_weight="1"
```

```
24            android:orientation="vertical">
25            <TextView
26                android:layout_width="wrap_content"
27                android:layout_height="wrap_content"
28                android:text="沉睡的海螺"
29                android:textColor="#000"
30                android:textSize="18sp" />
31            <TextView
32                android:layout_width="wrap_content"
33                android:layout_height="0dp"
34                android:layout_weight="1"
35                android:gravity="bottom"
36                android:text="2020年4月7日" />
37        </LinearLayout>
38        <Button
39            android:layout_width="0dp"
40            android:layout_height="wrap_content"
41            android:layout_gravity="center"
42            android:layout_weight="0.3"
43            android:text="关注" />
44    </LinearLayout>
45    <ImageView
46        android:layout_width="411dp"
47        android:layout_height="411dp"
48        android:scaleType="centerCrop"
49        android:src="@mipmap/img_pic" />
50    <TextView
51        android:layout_width="wrap_content"
52        android:layout_height="wrap_content"
53        android:layout_gravity="end"
54        android:layout_margin="5dp"
55        android:text="举报" />
56 </LinearLayout>
```

第 06 行 android:orientation="vertical"设置所有子控件以垂直线性布局排列。第 12 行 android: orientation="horizontal" 设置所有子控件以水平线性布局排列。第 23 行 android: layout_weight="1"设置父控件布局方向上的剩余空间分配给该控件的比例。第 42 行 android: layout_weight="0.3"设置父控件布局方向上的剩余空间分配给该控件的比例。第 53 行 android:layout_gravity="end"设置该标签在父控件内右侧对齐。

3.2 相对布局

3.2.1 RelativeLayout 控件

RelativeLayout 控件（android.widget.RelativeLayout）是相对排列布局的 UI 控件，即子控件以相对于父控件或相对于父控件中另一个子控件的相对位置进行放置的布局方式，是 android.view.ViewGroup 的子类，除继承的标签属性外，只有 2 个标签属性（如表 3-3 所示）。

其子控件在 RelativeLayout 控件中的定位是通过子控件属性进行设置的（如表 3-4 所示）。

表 3-3　RelativeLayout 控件的 XML 标签属性

属　　性	说　　明
android:gravity	设置控件内部的对齐方式
android:ignoreGravity	设置是否忽略 android:gravity 属性的设置

表 3-4　RelativeLayout 子控件的 XML 标签属性

属　　性	说　　明
android:layout_above	设置该控件位于指定 id 的控件上方，下边与指定 id 的控件上边对齐
android:layout_below	设置该控件位于指定 id 的控件下方，上边与指定 id 的控件下边对齐
android:layout_toStartOf	设置该控件位于指定 id 的控件左侧，右边与指定 id 的控件左边对齐
android:layout_toEndOf	设置该控件位于指定 id 的控件右侧，左边与指定 id 的控件右边对齐
android:layout_alignTop	设置该控件的上边和指定 id 的控件的上边对齐
android:layout_alignBottom	设置该控件的下边和指定 id 的控件的下边对齐
android:layout_alignLeft	设置该控件的左边和指定 id 的控件的左边对齐
android:layout_alignRight	设置该控件的右边和指定 id 的控件的右边对齐
android:layout_alignParentStart	设置该控件的左边和父控件的左边对齐
android:layout_alignParentTop	设置该控件的上边和父控件的上边对齐
android:layout_alignParentEnd	设置该控件的右边和父控件的右边对齐
android:layout_alignParentBottom	设置该控件的下边和父控件的下边对齐
android:layout_centerInParent	设置该控件位于父控件的中心位置
android:layout_centerHorizontal	设置该控件位于水平方向的中心位置
android:layout_centerVertical	设置该控件位于垂直方向的中心位置

3.2.2　实例工程：显示方位的相对布局

本实例演示了 9 种相对布局的效果。其中，"中""顶部""底部""左侧""右侧"是相对于父控件的相对布局，"上""下""左""右"是相对于"中"的相对布局（如图 3-2 所示）。

图 3-2　运行效果

1. 新建工程

新建一个"Empty Activity"工程，工程名称为"C0302"。

2. 主界面的布局

```xml
/res/layout/activity_main.xml
01  <?xml version="1.0" encoding="utf-8"?>
02  <RelativeLayout xmlns:android="http://schemas.android.com/apk/res/android"
03      android:layout_width="match_parent"
04      android:layout_height="match_parent">
05      <!--在RelativeLayout控件顶部显示-->
06      <TextView
07          android:layout_width="wrap_content"
08          android:layout_height="wrap_content"
09          android:layout_alignParentTop="true"
10          android:layout_centerHorizontal="true"
11          android:text="顶部"
12          android:textSize="30sp" />
13      <!--在RelativeLayout控件底部显示-->
14      <TextView
15          android:layout_width="wrap_content"
16          android:layout_height="wrap_content"
17          android:layout_alignParentBottom="true"
18          android:layout_centerHorizontal="true"
19          android:text="底部"
20          android:textSize="30sp" />
21      <!--在RelativeLayout控件左侧显示-->
22      <TextView
23          android:layout_width="wrap_content"
24          android:layout_height="wrap_content"
25          android:layout_alignParentStart="true"
26          android:layout_centerVertical="true"
27          android:text="左侧"
28          android:textSize="30sp" />
29      <!--在RelativeLayout控件右侧显示-->
30      <TextView
31          android:layout_width="wrap_content"
32          android:layout_height="wrap_content"
33          android:layout_alignParentEnd="true"
34          android:layout_centerVertical="true"
35          android:text="右侧"
36          android:textSize="30sp" />
37      <!--在RelativeLayout控件居中显示-->
38      <TextView
39          android:id="@+id/text_view"
40          android:layout_width="wrap_content"
41          android:layout_height="wrap_content"
42          android:layout_centerInParent="true"
43          android:text="中"
44          android:textSize="30sp" />
```

```
45          <!--显示在text_view上方-->
46          <TextView
47              android:layout_width="wrap_content"
48              android:layout_height="wrap_content"
49              android:layout_above="@id/text_view"
50              android:layout_centerHorizontal="true"
51              android:text="上"
52              android:textSize="30sp" />
53          <!--显示在text_view下方-->
54          <TextView
55              android:layout_width="wrap_content"
56              android:layout_height="wrap_content"
57              android:layout_below="@id/text_view"
58              android:layout_centerHorizontal="true"
59              android:text="下"
60              android:textSize="30sp" />
61          <!--显示在text_view左侧-->
62          <TextView
63              android:layout_width="wrap_content"
64              android:layout_height="wrap_content"
65              android:layout_centerVertical="true"
66              android:layout_toStartOf="@id/text_view"
67              android:text="左"
68              android:textSize="30sp" />
69          <!--显示在text_view右侧-->
70          <TextView
71              android:layout_width="wrap_content"
72              android:layout_height="wrap_content"
73              android:layout_centerVertical="true"
74              android:layout_toEndOf="@id/text_view"
75              android:text="右"
76              android:textSize="30sp" />
77      </RelativeLayout>
```

第 06~44 行的 5 个标签通过属性设置 RelativeLayout 控件内部的 5 种定位方式，android:layout_centerVertical="true" 设置左侧和右侧的垂直居中，android:layout_centerHorizontal="true"设置顶部和底部的垂直居中。第 46~76 行的 4 个标签通过属性设置与 text_view 的相对位置。

3.3 表格布局

3.3.1 TableLayout 控件

TableLayout 控件（android.widget.TableLayout）是使用表格形式布局的 UI 控件，是 android.widget.LinearLayout 的子类，除继承的标签属性外，还有 3 个标签属性（如表 3-5 所示）。TableLayout 控件还需要与 TableRow 控件配合使用，TableRow 控件是表格的行 UI 控件，其每一个子控件放置于一个单元格中，并可以通过子控件设置单元格的相关属性（如表 3-6 所示）。

表 3-5　TableLayout 控件的 XML 标签属性

属　　性	说　　明
android:collapseColumns	设置需要被隐藏的列序号
android:shrinkColumns	设置允许被收缩的列序号
android:stretchColumns	设置运行被拉伸的列序号

表 3-6　TableRow 子控件的 XML 标签属性

属　　性	说　　明
android:layout_column	设置跳过的单元格数量
android:layout_span	设置与当前单元格合并的单元格数量

3.3.2　实例工程：登录界面的表格视图

本实例演示了使用表格布局显示登录界面，"登录"按钮占用两个单元格，其余控件各占用一个单元格（如图 3-3 所示）。

图 3-3　运行效果

1．新建工程

新建一个"Empty Activity"工程，工程名称为"C0303"。

2．主界面的布局

```
/res/layout/activity_main.xml
01  <?xml version="1.0" encoding="utf-8"?>
02  <TableLayout xmlns:android="http://schemas.android.com/apk/res/android"
03      android:layout_width="300dp"
04      android:layout_height="wrap_content"
05      android:layout_gravity="center"
06      android:collapseColumns="2"
07      android:stretchColumns="1">
```

```
08      <!--第1行-->
09      <TableRow>
10          <!--第1列-->
11          <TextView
12              android:layout_width="wrap_content"
13              android:layout_height="wrap_content"
14              android:text="账号" />
15          <!--第2列-->
16          <EditText
17              android:layout_width="wrap_content"
18              android:layout_height="wrap_content"
19              android:ems="10"
20              android:inputType="textPersonName" />
21      </TableRow>
22      <!--第2行-->
23      <TableRow>
24          <!--第1列-->
25          <TextView
26              android:layout_width="wrap_content"
27              android:layout_height="wrap_content"
28              android:text="密码" />
29          <!--第2列-->
30          <EditText
31              android:layout_width="wrap_content"
32              android:layout_height="wrap_content"
33              android:ems="10"
34              android:inputType="textPassword" />
35          <!--第3列-->
36          <TextView
37              android:layout_width="wrap_content"
38              android:layout_height="wrap_content"
39              android:text="忘记密码" />
40      </TableRow>
41      <!--第3行-->
42      <TableRow>
43          <!--第2列-->
44          <TextView
45              android:id="@+id/textView4"
46              android:layout_width="wrap_content"
47              android:layout_height="wrap_content"
48              android:layout_column="1"
49              android:gravity="end"
50              android:text="注册账号" />
51      </TableRow>
52      <!--第4行-->
53      <TableRow>
54          <!--第1列和第2列合并-->
55          <Button
56              android:layout_width="wrap_content"
57              android:layout_height="wrap_content"
```

```
58             android:layout_span="2"
59             android:text="登录" />
60     </TableRow>
61 </TableLayout>
```

第 06 行 android:collapseColumns="2"设置表格布局的第 3 列隐藏不显示。第 07 行 android:stretchColumns="1"设置表格布局的第 2 列可以被拉伸。第 48 行 android:layout_column="1"设置跳过当前行的前一个单元格。第 58 行 android:layout_span="2"设置表格布局的第 4 行第 1 列和第 2 列的单元格合并为一个单元格。

3.4 网格布局

3.4.1 GridLayout 控件

GridLayout 控件（android.widget.GridLayout）是网格形式排列布局的 UI 控件，是 android.view.ViewGroup 的子类。GridLayout 控件的常用 XML 标签属性如表 3-7 所示。

表 3-7　GridLayout 控件的常用 XML 标签属性

属　　性	说　　明
android:alignmentMode	设置对齐模式，可选值为 alignBounds 和 alignMargins
android:columnCount	设置最大的列数
android:orientation	设置子控件排列顺序的方向，可选值包括 horizontal 和 vertical
android:rowCount	设置最大的行数
android:useDefaultMargins	设置是否使用默认边距

3.4.2 实例工程：模仿计算器界面的网格布局

本实例演示了使用网格布局显示计算器的界面，第一行的控件占用 4 列网格，第二行的控件各占用 2 列网格（如图 3-4 所示）。

图 3-4　运行效果

1. 新建工程

新建一个"Empty Activity"工程，工程名称为"C0304"。

2. 主界面的布局

```xml
/res/layout/activity_main.xml
01  <?xml version="1.0" encoding="utf-8"?>
02  <GridLayout xmlns:android="http://schemas.android.com/apk/res/android"
03      android:layout_width="wrap_content"
04      android:layout_height="wrap_content"
05      android:layout_gravity="center"
06      android:background="#eeeeee"
07      android:columnCount="4"
08      android:rowCount="6">
09      <TextView
10          android:layout_columnSpan="4"
11          android:layout_gravity="fill"
12          android:gravity="right"
13          android:layout_marginLeft="5dp"
14          android:layout_marginRight="5dp"
15          android:paddingRight="10dp"
16          android:background="#999999"
17          android:text="0"
18          android:textSize="50sp" />
19      <Button
20          android:layout_columnSpan="2"
21          android:layout_gravity="fill"
22          android:text="回退" />
23      <Button
24          android:layout_columnSpan="2"
25          android:layout_gravity="fill"
26          android:text="清空" />
27      <Button android:text="+" />
28      <Button android:text="1" />
29      <Button android:text="2" />
30      <Button android:text="3" />
31      <Button android:text="-" />
32      <Button android:text="4" />
33      <Button android:text="5" />
34      <Button android:text="6" />
35      <Button android:text="*" />
36      <Button android:text="7" />
37      <Button android:text="8" />
38      <Button android:text="9" />
39      <Button android:text="/" />
40      <Button android:text="." />
41      <Button android:text="0" />
42      <Button android:text="=" />
43  </GridLayout>
```

第07行 android:columnCount="4" 设置网格布局分为4列。第08行 android:rowCount="6" 设置网格布局分为 6 行。第 10 行 android:layout_columnSpan="4" 设置 TextView 控件占用 4 列网格。第 19~26 行 android:layout_columnSpan="2" 设置 Button 控件占用 2 列网格，android:layout_gravity="fill" 设置完全充满父控件水平和垂直方向的空间。

3.5 帧 布 局

3.5.1 FrameLayout 控件

FrameLayout 控件（android.widget.FrameLayout）是依次堆叠形式排列布局的 UI 控件，是 android.view.ViewGroup 的子类，后添加的子控件置于上层，除继承的标签属性外，还有 2 个标签属性（如表 3-8 所示）。

表 3-8 FrameLayout 控件的 XML 标签属性

属性	说明
android:foregroundGravity	设置前景图像显示的位置
android:measureAllChildren	设置是否对子控件进行测量，默认值为 true

3.5.2 实例工程：分层显示图像的帧布局

本实例演示了使用帧布局叠加显示图像的效果，文字和汽车图像包含透明区域，透明区域直接显示出背景图像（如图 3-5 所示）。

图 3-5 运行效果

1. 新建工程并导入素材

新建一个"Empty Activity"工程，工程名称为"C0305"。选择"/res/mipmap"资源文件夹，将素材图像（素材文件夹路径为"/素材/C0305"）粘贴或拖曳到该文件夹中。

2．主界面的布局

```
/res/layout/activity_main.xml
01  <?xml version="1.0" encoding="utf-8"?>
02  <FrameLayout xmlns:android="http://schemas.android.com/apk/res/android"
03      android:layout_width="wrap_content"
04      android:layout_height="wrap_content">
05      <!--背景-->
06      <ImageView
07          android:layout_width="wrap_content"
08          android:layout_height="wrap_content"
09          android:scaleType="fitXY"
10          android:src="@mipmap/img_bg" />
11      <!--汽车-->
12      <ImageView
13          android:layout_width="wrap_content"
14          android:layout_height="wrap_content"
15          android:src="@mipmap/img_car" />
16      <!--标题-->
17      <ImageView
18          android:layout_width="wrap_content"
19          android:layout_height="wrap_content"
20          android:src="@mipmap/img_title" />
21  </FrameLayout>
```

第 06～20 行添加 3 个<ImageView>标签，自下而上进行排列。图像采用带透明度通道的 png 格式，汽车和标题的图像透明区域直接显示出背景图像。

3.6 约束布局

3.6.1 ConstraintLayout 控件

ConstraintLayout 控件（androidx.constraintlayout.widget.ConstraintLayout）是依次堆叠形式排列布局的 UI 控件，其子控件通过标签属性进行约束（如表 3-9 所示）。ConstraintLayout 与 RelativeLayout 类似，但嵌套层级更少且更加灵活。定位标签的属性值为 "parent"，是指子控件相对于 ConstraintLayout 控件的位置。定位标签的属性值为其他子控件的 id，是指子控件相对于其他子控件的位置。

表 3-9 ConstraintLayout 子控件的常用 XML 标签属性

属性	说明
app:layout_constraintStart_toStartOf	设置左侧与指定控件的左侧约束
app:layout_constraintStart_toEndOf	设置左侧与指定控件的右侧约束
app:layout_constraintEnd_toStartOf	设置右侧与指定控件的左侧约束
app:layout_constraintEnd_toEndOf	设置右侧与指定控件的右侧约束
app:layout_constraintTop_toTopOf	设置顶端与指定控件的顶端约束
app:layout_constraintTop_toBottomOf	设置顶端与指定控件的底端约束

续表

属　性	说　明
app:layout_constraintBottom_toTopOf	设置底端与指定控件的顶端约束
app:layout_constraintBottom_toBottomOf	设置底端与指定控件的底端约束
app:layout_constraintDimensionRatio	设置宽度和高度的比例。高度和宽度至少有一个值应设置为"0dp"，在运行时根据该属性值自动计算设置为"0dp"的尺寸；如果都为"0dp"，在运行时根据所有约束条件和该属性值自动计算最大尺寸
app:layout_constraintCircleAngle	设置约束角度
app:layout_constraintCircleRadius	设置约束角度的半径

 提示：AndroidX

AndroidX 是 Google 公司从 Android 9.0（API level 28）开始推荐使用的新支持库，具有实时更新、无须经常修改版本的优点。例如，老版本的 android.support.v4.app 支持库的 v4 表示 Android API 版本号，提供的 API 会向下兼容到 Android 1.6（使用 API level 4.0）。原有的支持库虽然被保留可以继续使用，但从 Android Studio 3.4.2 开始，新建项目默认使用 AndroidX 架构了。原有项目可以使用菜单【Refactor】→【Migrate to AndroidX】命令进行迁移。

3.6.2　实例工程：模仿朋友圈顶部的约束布局

本实例演示了使用约束布局模仿朋友圈顶部的效果（如图 3-6 所示）。背景图像与其父控件进行约束，头像图像与背景图像进行约束，昵称文本与头像图像进行约束。

1．新建工程并导入素材

新建一个 "Empty Activity" 工程，工程名称为 "C0306"。选择 "/res/mipmap" 资源文件夹，将素材图像（素材文件夹路径为 "/素材/C0306"）粘贴或拖曳到该文件夹中。

2．主界面的布局

图 3-6　运行效果

```
/res/layout/activity_main.xml
01  <?xml version="1.0" encoding="utf-8"?>
02  <androidx.constraintlayout.widget.ConstraintLayout
03      xmlns:android="http://schemas.android.com/apk/res/android"
04      xmlns:app="http://schemas.android.com/apk/res-auto"
05      android:layout_width="match_parent"
06      android:layout_height="match_parent">
07      <!--背景图像-->
08      <ImageView
09          android:id="@+id/image_view_bg"
10          android:layout_width="wrap_content"
11          android:layout_height="0dp"
```

```
12            app:layout_constraintDimensionRatio="1.5"
13            app:layout_constraintStart_toStartOf="parent"
14            app:layout_constraintTop_toTopOf="parent"
15            app:srcCompat="@mipmap/img_bg" />
16       <!--头像图像-->
17       <ImageView
18            android:id="@+id/image_view_face"
19            android:layout_width="80dp"
20            android:layout_height="80dp"
21            android:layout_marginTop="212dp"
22            android:layout_marginEnd="30dp"
23            app:layout_constraintEnd_toEndOf="@+id/image_view_bg"
24            app:layout_constraintTop_toTopOf="@+id/image_view_bg"
25            app:srcCompat="@mipmap/img_face" />
26       <!--昵称-->
27       <TextView
28            android:layout_width="wrap_content"
29            android:layout_height="wrap_content"
30            android:layout_marginTop="25dp"
31            android:layout_marginEnd="20dp"
32            android:text="沉睡的海螺"
33            android:textColor="#FFFFFF"
34            android:textSize="18sp"
35            app:layout_constraintEnd_toStartOf="@+id/image_view_face"
36            app:layout_constraintTop_toTopOf="@+id/image_view_face" />
37  </androidx.constraintlayout.widget.ConstraintLayout>
```

第 12～14 行 app:layout_constraintDimensionRatio="1.5"设置<ImageView>标签宽度和高度的约束比例，app:layout_constraintStart_toStartOf="parent"设置<ImageView>标签右侧与父控件的右侧进行约束，app:layout_constraintTop_toTopOf="parent"设置<ImageView>标签顶端与父控件的顶端进行约束。第 21～24 行设置<ImageView>标签与 image_view_bg 对象的约束关系，通过 android:layout_marginTop="212dp"设置与 image_view_bg 对象顶端的距离，通过 android:layout_marginEnd="30dp"设置与 image_view_bg 对象顶端的距离。第 35、36 行设置<TextView>标签与 image_view_face 对象的约束关系。

 提示：AbsoluteLayout（绝对布局）

AbsoluteLayout 在 API level 3 的时候就已经不推荐使用了，其子控件使用 android:layout_x 和 android:layout_y 设置坐标。对于屏幕适配的情况，绝对布局还需要进行转换，此时可以考虑使用 ConstraintLayout（约束布局）。

3.7 习 题

1．实现注册界面的布局效果，至少包含用户名、密码、忘记密码、登录等控件。
2．实现手机遥控器的布局效果，至少包含数字 0～9、十字形布局的频道和音量控件。
3．实现类似淘宝或京东首页商品类别的布局效果，至少包含 2 行，每行 4 个商品类型。

第 4 章　UI 控件与数据适配

有一类 UI 控件单独使用无法实现其应有的功能，需要适配器或与其他控件配合使用。这类 UI 控件中有些通过适配器会将数据匹配后显示出来，有些可以作为其他 UI 控件的容器。这类 UI 控件的主要特点是功能和内容分离，降低了耦合性，特别适合可变长度的内容呈现。

4.1　数据适配原理

有一类 UI 控件视图（View）的内部包含子视图（ItemView），每个子视图（ItemView）对应一组子数据（ItemData）。但是，不直接通过 UI 控件标签属性或类方法设置所需的数据，而是通过适配器（Adapter）将子视图与对应的子数据关联起来（如图 4-1 所示）。当子数据包含的数据项较多时，如果使用数组存储和传递子数据，就需要几个甚至十几个数组，不利于管理和使用。这种情况下，推荐将子数据存储在模型（Model）中，然后使用模型通过适配器与子视图进行关联，这种模式属于 MVC 模式。

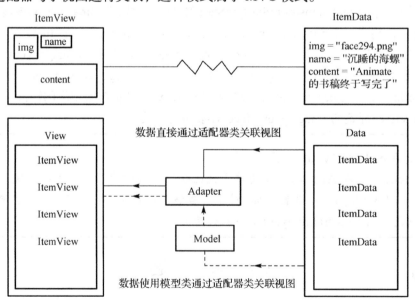

图 4-1　数据适配原理

提示:MVC 模式

MVC 是模型(Model)、视图(View)、控制器(Controller)的缩写。MVC 模式是将业务逻辑、数据、界面分离来组织代码的。

4.2 列表视图

4.2.1 ListView 控件

ListView 控件(android.widget.ListView)是滚动显示列表信息的 UI 控件,是 android.widget.AbsListView 的子类,通过适配器将数据与列表项相匹配。除继承的标签属性外,还有 5 个标签属性(如表 4-1 所示),而类方法可以提供设置适配器、顶部视图、底部视图、滚动等更多功能(如表 4-2 所示)。

表 4-1 ListView 控件的 XML 标签属性

属 性	说 明
android:divider	设置分割线的图像资源
android:dividerHeight	设置分割线的高度
android:entries	设置填充控件的数组资源引用
android:footerDividersEnabled	设置是否显示与底部视图之间的分割线
android:headerDividersEnabled	设置是否显示与顶部视图之间的分割线

表 4-2 ListView 类的常用方法

类型和修饰符	方 法
void	addFooterView(View v) 添加底部视图
void	addHeaderView(View v) 添加顶部视图
ListAdapter	getAdapter() 获取适配器对象
View	getChildAt(int index) 获取指定位置的视图
int	getMaxScrollAmount() 获取响应箭头事件滚动的最大值
boolean	removeFooterView(View v) 移除底部视图
boolean	removeHeaderView(View v) 移除顶部视图
void	setOnItemClickListener(AdapterView.OnItemClickListener listener) 设置单击列表项的监听器
void	setAdapter(ListAdapter adapter) 设置适配器对象
void	setSelector(Drawable sel) 设置被选中项的背景
void	setSelection(int position) 设置被选择的列表项位置

续表

类型和修饰符	方法
void	setSelectionAfterHeaderView() 选中顶部视图下方的第一个列表项
void	smoothScrollByOffset(int offset) 平滑地滚动到指定的适配器偏移位置，offset 是与当前位置的偏移量
void	smoothScrollToPosition(int position) 平滑地滚动到指定的适配器位置

简单数据的列表项适配可以使用 ArrayAdapter 类或 SimpleAdapter 类。自定义的列表项适配需要创建 BaseAdpater 类的子类和自定义列表视图。

提示：RecyclerView 控件

如果想实现水平滚动或瀑布流的效果，需要使用官方扩展库提供的 RecyclerView 控件，在 Android Studio 的 Palette 面板中可以下载。

4.2.2 实例工程：简单数据的列表视图

本实例演示了从资源文件中获取字符串数组显示到列表视图中，并为列表视图添加顶部视图，单击列表项后显示相应的提示信息（如图 4-2 所示）。

1. 新建工程

新建一个"Empty Activity"工程，工程名称为"C0401"。

2. 列表项的布局

在"layout"资源文件夹中，新建"list_view_header.xml"文件，该文件用于保存 ListView 控件的顶部视图。

图 4-2 运行效果

```
/res/layout/list_view_header.xml
01  <?xml version="1.0" encoding="utf-8"?>
02  <LinearLayout xmlns:android="http://schemas.android.com/apk/res/android"
03      android:layout_width="match_parent"
04      android:layout_height="match_parent"
05      android:orientation="vertical">
06      <TextView
07          android:layout_width="match_parent"
08          android:layout_height="wrap_content"
09          android:layout_margin="15dp"
10          android:gravity="center"
11          android:text="选择电影类型"
12          android:textSize="24dp" />
13  </LinearLayout>
```

第 06～12 行<TextView>标签显示 ListView 控件顶部视图中的标题文字，居中显示，外边距设置为 "15dp"。

3. 主界面的布局

```
/res/layout/activity_main.xml
01  <?xml version="1.0" encoding="utf-8"?>
02  <LinearLayout xmlns:android="http://schemas.android.com/apk/res/android"
03      android:layout_width="match_parent"
04      android:layout_height="match_parent"
05      android:orientation="vertical">
06      <ListView
07          android:id="@+id/list_view"
08          android:layout_width="match_parent"
09          android:layout_height="match_parent" />
10  </LinearLayout>
```

第 06～09 行<ListView>标签的属性并不能设置顶部视图，需要通过类方法进行设置。

4. 字符串资源

```
/res/values/strings.xml
01  <resources>
02      <string name="app_name">C0401</string>
03      <string-array name="movie_type_array">
04          <item>剧情</item>
05          <item>喜剧</item>
06          <item>爱情</item>
07          <item>传记</item>
08          <item>历史</item>
09          <item>科幻</item>
10          <item>奇幻</item>
11          <item>悬疑</item>
12          <item>犯罪</item>
13          <item>武侠</item>
14          <item>动画</item>
15      </string-array>
16  </resources>
```

第 03 行<string-array>标签保存字符串数组，数组名称属性设置为 "movie_type_array"，该标签的资源 id 为 R.array.movie_type_array。第 04～14 行<item>标签表示数组项，保存电影类型的名称。

5. 主界面的 Activity

```
/java/com/vt/c0401/MainActivity.java
13  public class MainActivity extends AppCompatActivity {
14      private ListView mListView;
15      @Override
16      protected void onCreate(Bundle savedInstanceState) {
17          super.onCreate(savedInstanceState);
18          setContentView(R.layout.activity_main);
```

19	`        ArrayAdapter<CharSequence> resAdapter = ArrayAdapter.`*`createFromResource`*`(`**`this`**`,R.array.`***`movie_type_array`***`,android.R.layout.`***`simple_expandable_list_item_1`***`);//movie_type_array`数组资源适配
20	`        View  headerView = LayoutInflater.`*`from`*`(MainActivity.`**`this`**`).inflate(R.layout.`***`list_view_header`***`, null);//`顶部视图
21	`        `**`mListView`**` = findViewById(R.id.`*`list_view`*`);//`获取<ListView>标签
22	`        `**`mListView`**`.setAdapter(resAdapter);//`设置数据适配器
23	`        `**`mListView`**`.addHeaderView(headerView);//`添加顶部视图
24	`        //`添加单击列表项事件监听器
25	`        `**`mListView`**`.setOnItemClickListener(`**`new`**` AdapterView.OnItemClickListener() {`
26	`            @Override`
27	`            public void onItemClick(AdapterView<?> parent, View view, `**`int`**` position, `**`long`**` id) {`
28	`                Toast.`*`makeText`*`(MainActivity.`**`this`**`,((TextView)`**`mListView`**`.getChildAt(position)).getText(), Toast.`***`LENGTH_SHORT`***`).show();`
29	`            }`
30	`        });`
31	`    }`
32	`}`

第 19 行使用 ArrayAdapter.createFromResource 静态方法初始化一个 ArrayAdapter 实例；android.R.layout.simple_expandable_list_item_1 是 API 自带的列表项视图，内部只包含一个 TextView 控件。第 20 行通过 LayoutInflater 类的静态方法获取顶部视图。第 21～31 行先从布局文件中获取 ListView 控件，设置适配器并添加顶部视图，然后为列表项添加单击列表项事件监听器。

4.2.3 实例工程：带缓存的自定义列表视图

本实例演示了常用的自定义列表视图（如图 4-3 所示），列表项的布局保存在布局文件中，列表项数据通过模型类传递给适配器，为了提升运行效率，还进行了缓存处理。

1．新建工程并导入素材

新建一个"Empty Activity"工程，工程名称为"C0402"。选择"/res/mipmap"资源文件夹，将素材图像（素材文件夹路径为"/素材/C0402"）粘贴或拖曳到该文件夹中。

2．照片模型

在"/java/com/vt/c0402"文件夹中，新建"PhotoModel.java"文件，用于构建照片及其相关数据的模型类。

图 4-3 运行效果

	`/java/com/vt/c0402/PhotoModel.java`
03	`public class PhotoModel {`
04	`    public int photoResId;//`图像资源 id
05	`    public int hits;//`图像单击次数
06	`    public PhotoModel(int photoResId, int hits) {`
07	`        this.photoResId = photoResId;`
08	`        this.hits = hits;`

| 09 | } |
| 10 | } |

第 04、05 行是两个公有属性，存储照片的资源 id 和照片单击的次数。第 06～09 行是构造方法，通过其参数初始化公有属性。

3．列表项的布局

在"/res/layout"文件夹中，新建"list_view_item.xml"文件，用于保存列表项的布局，将 PhotoModel 类实例提供的数据处理后呈现在 ListView 控件中。

/res/layout/list_view_item.xml

```xml
01  <?xml version="1.0" encoding="utf-8"?>
02  <LinearLayout xmlns:android="http://schemas.android.com/apk/res/android"
03      xmlns:tools="http://schemas.android.com/tools"
04      android:layout_width="match_parent"
05      android:layout_height="match_parent"
06      android:orientation="vertical"
07      android:showDividers="middle">
08      <TextView
09          android:id="@+id/text_view_order"
10          android:layout_width="match_parent"
11          android:layout_height="wrap_content"
12          android:layout_margin="10dp"
13          android:gravity="center"
14          android:textSize="18sp"
15          tools:text="NO.1" />
16      <ImageView
17          android:id="@+id/image_view_photo"
18          android:layout_width="match_parent"
19          android:layout_height="wrap_content"
20          android:adjustViewBounds="true"
21          tools:srcCompat="@tools:sample/avatars" />
22      <TextView
23          android:id="@+id/text_view_hits"
24          android:layout_width="match_parent"
25          android:layout_height="wrap_content"
26          android:layout_margin="5dp"
27          android:gravity="end"
28          tools:text="单击量: 201" />
29  </LinearLayout>
```

第 16～21 行<ImageView>标签显示照片，srcCompat 属性与 src 属性相比可以保证版本的兼容性。命名空间使用 tools 替换 android，用于在 IDE 中预览渲染。@tools:sample/avatars 是自带的样例图像，可在 IDE 中预览效果。

4．列表视图的适配器

在"/java/com/vt/c0402"文件夹中，新建"PhotoListAdapter.java"文件，用于创建 ListView 控件的自定义适配器类——PhotoListAdapter 类，该类继承了 BaseAdapter 类。

```
/java/com/vt/c0402/PhotoListAdapter.java
14  public class PhotoListAdapter extends BaseAdapter {
15      private Context mContext;
16      private List<PhotoModel> mPhotoModel;
17      private ViewHolder mViewHolder;
18      private int mConvertViewCount;
19      public PhotoListAdapter(Context mContext, List<PhotoModel> mPhotoModel) {
20          this.mContext = mContext;
21          this.mPhotoModel = mPhotoModel;
22          this.mConvertViewCount = 0;
23      }
24      //获取列表项总数
25      @Override
26      public int getCount() {
27          return mPhotoModel.size();
28      }
29      //获取列表项对象
30      @Override
31      public Object getItem(int position) {
32          return mPhotoModel;
33      }
34      //获取列表项 id
35      @Override
36      public long getItemId(int position) {
37          return position;
38      }
39      //获取列表项视图
40      @Override
41      public View getView(int position, View convertView, ViewGroup parent) {
42          if (convertView == null) {//convertView 未实例化时
43              convertView = LayoutInflater.from(mContext).inflate(R.layout.list_view_item, parent, false);
44              mConvertViewCount ++;
45              //实例化 ViewHolder 缓存类对象
46              mViewHolder = new ViewHolder();
47              mViewHolder.orderTextView = convertView.findViewById(R.id.text_view_order);
48              mViewHolder.photoImageView = convertView.findViewById(R.id.image_view_photo);
49              mViewHolder.hitsTextView = convertView.findViewById(R.id.text_view_hits);
50              convertView.setTag(mViewHolder);//holder 缓存到 convertView
51              convertView.setId(mConvertViewCount);
52              Log.i("getView","当前位置 position: "+position+", 初始化 convertView 的 id: " +convertView.getId()+". ");
53          } else {//convertView 实例化时
54              mViewHolder = (ViewHolder) convertView.getTag();//从 convertView 获取 holder 缓存
55              Log.i("getView","当前位置 position: "+position+", 重用 convertView 的 id: " +convertView.getId()+". ");
56          }
57          mViewHolder.orderTextView.setText("No." + (position + 1));
58          mViewHolder.photoImageView.setImageResource(mPhotoModel.get(position).photoResId);
59          mViewHolder.hitsTextView.setText("单击量:" + mPhotoModel.get(position).hits);
```

```
60              mViewHolder.photoImageView.setId(position);
61              mViewHolder.photoImageView.setOnClickListener(new View.OnClickListener() {
62                  public void onClick(View v) {
63                      mPhotoModel.get(v.getId()).hits++;
64                      notifyDataSetChanged();//更新视图数据
65                      Toast.makeText(mContext, "单击No." + ((int) v.getId() + 1), Toast.LENGTH_SHORT).show();
66                  }
67              });
68              return convertView;
69          }
70          //ViewHolder缓存类
71          class ViewHolder {
72              TextView orderTextView;
73              ImageView photoImageView;
74              TextView hitsTextView;
75          }
76      }
```

第 41 行实现的 getView(int position, View convertView, ViewGroup parent)方法是 Adapter 接口的抽象方法，初始化、滚动父视图再调用列表项或执行 notifyDataSetChanged() 时被调用；position 是列表项的位置，convertView 是列表项的视图，parent 是列表项的父视图。第 42～56 行判断 convertView 是否为空，等于 null 时对其进行初始化并缓存到 convertView 的标签对象中，不等于 null 时直接从 convertView 的标签对象中获取。第 61～67 行是 photoImageView 对象的单击事件监听器，notifyDataSetChanged()语句能够立即根据数据的改变更新 ListView 控件的列表项视图。第 71～75 行是一个内部类，包含与列表项视图中控件相对应类型的三个属性，用于缓存列表项视图中的控件。

5．主界面的布局

```
/res/layout/activity_main.xml
01  <?xml version="1.0" encoding="utf-8"?>
02  <LinearLayout xmlns:android="http://schemas.android.com/apk/res/android"
03      xmlns:tools="http://schemas.android.com/tools"
04      android:layout_width="match_parent"
05      android:layout_height="match_parent"
06      tools:context=".MainActivity">
07      <ListView
08          android:id="@+id/list_view_photo"
09          android:layout_width="match_parent"
10          android:layout_height="match_parent" />
11  </LinearLayout>
```

第 07～10 行<ListView>标签显示图像序号、图像和图像单击次数的列表。

6．主界面的 Activity

```
/java/com/vt/c0402/MainActivity.java
10  public class MainActivity extends AppCompatActivity {
11      private Context mContext;
12      private List<PhotoModel> mPhotoModel;
13      private PhotoListAdapter mAdapter;
```

```
14      private ListView photoListView;
15      @Override
16      protected void onCreate(Bundle savedInstanceState) {
17          super.onCreate(savedInstanceState);
18          setContentView(R.layout.activity_main);
19          mContext = MainActivity.this;
20          mPhotoModel = new ArrayList<>();
21          mPhotoModel.add(new PhotoModel(R.mipmap.img_pic01, 203));
22          mPhotoModel.add(new PhotoModel(R.mipmap.img_pic02, 44));
23          mPhotoModel.add(new PhotoModel(R.mipmap.img_pic03, 311));
24          mPhotoModel.add(new PhotoModel(R.mipmap.img_pic04, 97));
25          mPhotoModel.add(new PhotoModel(R.mipmap.img_pic05, 462));
26          mPhotoModel.add(new PhotoModel(R.mipmap.img_pic06, 117));
27          mPhotoModel.add(new PhotoModel(R.mipmap.img_pic07, 187));
28          mPhotoModel.add(new PhotoModel(R.mipmap.img_pic08, 87));
29          mAdapter = new PhotoListAdapter(mContext,mPhotoModel);
30          photoListView = findViewById(R.id.list_view_photo);
31          photoListView.setAdapter(mAdapter);
32      }
33  }
```

第 21~28 行实例化 List<PhotoModel>对象，添加 8 个 PhotoModel 实例，PhotoModel 实例保存图像的资源 id 和单击次数。第 29 行将上下文和保存 PhotoModel 实例的 List<PhotoModel>对象作为参数，实例化自定义的适配器——PhotoListAdapter。第 31 行 ListView 控件设置适配器。

7．测试运行

运行程序后，拖动 ListView 控件直至显示第 6 个列表项视图，ListView 控件回到顶部，可以在"Logcat"窗口中观察输出的日志信息（如图 4-4 所示）。在 ListView 控件中，列表项同时最多可以显示 3 个，因此初始化的 convertView 对象只有 3 个，滚动 ListView 控件时 3 个 convertView 对象被循环重用。

图 4-4 "Logcat"窗口输出的日志信息

4.3 网格视图

4.3.1 GridView 控件

GridView 控件（android.widget.GridView）是以滚动网格形式显示信息的 UI 控件，是

android.widget.AbsListView 的子类，通过适配器将数据与网格项相匹配。除继承的标签属性外，还有 6 个标签属性（如表 4-3 所示），而类方法可以提供设置适配器、滚动等更多功能（如表 4-4 所示）。

表 4-3　GridView 控件的 XML 标签属性

属　　性	说　　明
android:columnWidth	设置列宽度
android:gravity	设置对齐方式
android:horizontalSpacing	设置水平方向单元格的间距
android:numColumns	设置列数
android:stretchMode	设置拉伸模式，可选值包括 none（不拉伸）、spacingWidth（拉伸单元格的间距）、columnWidth（拉伸列宽度）和 spacingWidthUniform（均匀拉伸单元格的间距）
android:verticalSpacing	设置垂直方向单元格的间距

表 4-4　GridView 类的常用方法

类型和修饰符	方　　法
void	setAdapter(ListAdapter adapter) 设置适配器对象
void	setOnItemClickListener(AdapterView.OnItemClickListener listener) 设置单击列表项的监听器
void	setSelector(Drawable sel) 设置被选中项的背景
void	setSelection(int position) 设置被选择的列表项位置
void	smoothScrollByOffset(int offset) 平滑地滚动到指定的适配器偏移位置，offset 是与当前位置的偏移量
void	smoothScrollToPosition(int position) 平滑地滚动到指定的适配器位置

4.3.2　实例工程：显示商品类别的网格项视图

本实例演示了使用网格项视图分两行排列商品类别，每个类别都配有文字和图像，单击后显示相应的提示信息（如图 4-5 所示）。

1．新建工程并导入素材

新建一个"Empty Activity"工程，工程名称为"C0403"。选择"/res/mipmap"资源文件夹，将素材图像（素材文件夹路径为"/素材/C0403"）粘贴或拖曳到该文件夹中。

2．商品分类的模型

在"/java/com/vt/c0403"文件夹中，新建"CategoryModel.java"文件，用于创建商品分类的模型。

图 4-5　运行效果

```
/java/com/vt/c0403/CategoryModel.java
03  public class CategoryModel {
04      public int ico;
05      public String name;
06      public CategoryModel(int ico, String name) {
07          this.ico = ico;
08          this.name = name;
09      }
10  }
```

第 04、05 行是两个公有属性，存储商品类别图标的资源 id 和商品类别名称。第 06~09 行是构造方法，通过其参数初始化公有属性。

3．网格项视图

在"/res/layout"文件夹中，新建"grid_view_item.xml"文件，用于创建 GirdView 控件的网格项视图。

```
/res/layout/grid_view_item.xml
01  <?xml version="1.0" encoding="utf-8"?>
02  <RelativeLayout xmlns:android="http://schemas.android.com/apk/res/android"
03      xmlns:tools="http://schemas.android.com/tools"
04      android:layout_width="wrap_content"
05      android:layout_height="wrap_content">
06      <ImageView
07          android:id="@+id/image_view_ico"
08          android:layout_width="64dp"
09          android:layout_height="64dp"
10          android:layout_centerHorizontal="true"
11          android:adjustViewBounds="true"
12          tools:srcCompat="@tools:sample/avatars" />
13      <TextView
14          android:id="@+id/text_view_name"
15          android:layout_width="wrap_content"
16          android:layout_height="wrap_content"
17          android:layout_below="@id/image_view_ico"
18          android:layout_centerHorizontal="true"
19          android:textSize="18sp"
20          tools:text="类别名称" />
21  </RelativeLayout>
```

第 06~12 行<ImageView>标签显示商品类别的图像，使用自带的样例图像进行占位，命名空间使用 tools 替换 android，可以在 IDE 中预览渲染，srcCompat 属性替换 src 属性可以保证版本的兼容性。

4．网格项视图的适配器

在"/java/com/vt/c0403"文件夹中，新建"CategoryAdapter.java"文件，用于创建 GridView 控件的自定义适配器类——CategoryAdapter 类，该类继承了 BaseAdapter 类。

```
/java/com/vt/c0403/CategoryAdapter.java
13  public class CategoryAdapter extends BaseAdapter {
14      private Context mContext;
15      private List<CategoryModel> mCategoryModel;
16      private ViewHolder mViewHolder;
17      private int mConvertViewCount;
18      CategoryAdapter(Context mContext, List<CategoryModel> mCategoryModel) {
19          this.mContext = mContext;
20          this.mCategoryModel = mCategoryModel;
21      }
22      //获取网格项总数
23      @Override
24      public int getCount() {
25          return mCategoryModel.size();
26      }
27      //获取网格项对象
28      @Override
29      public Object getItem(int position) {
30          return mCategoryModel;
31      }
32      //获取网格项 id
33      @Override
34      public long getItemId(int position) {
35          return position;
36      }
37      //获取网格项视图
38      @Override
39      public View getView(int position, View convertView, ViewGroup parent) {
40          if (convertView == null) {//convertView 未实例化时
41              convertView = LayoutInflater.from(mContext).inflate(R.layout.grid_view_item, parent, false);
42              mConvertViewCount++;
43              //实例化 ViewHolder 缓存类对象
44              mViewHolder = new ViewHolder();
45              mViewHolder.icoImageView = convertView.findViewById(R.id.image_view_ico);
46              mViewHolder.nameTextView = convertView.findViewById(R.id.text_view_name);
47              convertView.setTag(mViewHolder);//holder 缓存到 convertView 中
48              convertView.setId(mConvertViewCount);
49              Log.i("getView", "当前位置 position: " + position + ", 初始化 convertView 的 id: " + convertView.getId() + ". ");
50          } else {//convertView 实例化时
51              mViewHolder = (ViewHolder) convertView.getTag();//从 convertView 获取 holder 缓存
52              Log.i("getView", "当前位置 position: " + position + ", 重用 convertView 的 id: " + convertView.getId() + ". ");
53          }
54
55          mViewHolder.icoImageView.setImageResource(mCategoryModel.get(position).ico);
56          mViewHolder.nameTextView.setText(mCategoryModel.get(position).name);
57          return convertView;
58      }
```

```
59        //ViewHolder缓存类
60        class ViewHolder {
61            ImageView icoImageView;
62            TextView nameTextView;
63        }
64    }
```

第 39 行实现的 getView(int position, View convertView, ViewGroup parent)方法是 Adapter 接口的抽象方法，返回网格项显示的视图。第 40～53 行判断 convertView 是否为空，等于 null 时对其进行初始化并缓存到 convertView 的标签对象中，不等于 null 时直接从 convertView 的标签对象中获取。第 60～63 行是一个内部类，包含与网格项视图中控件相对应类型的两个属性，缓存网格项视图中的控件。

5．主视图的布局

```
/res/layout/activity_main.xml
01  <RelativeLayout xmlns:android="http://schemas.android.com/apk/res/android"
02      android:layout_width="match_parent"
03      android:layout_height="match_parent"
04      android:paddingTop="25dp">
05      <GridView
06          android:id="@+id/grid_view"
07          android:layout_width="match_parent"
08          android:layout_height="wrap_content"
09          android:columnWidth="80dp"
10          android:gravity="center_horizontal"
11          android:numColumns="4"
12          android:stretchMode="spacingWidthUniform"
13          android:verticalSpacing="10dp" />
14  </RelativeLayout>
```

第 05～13 行<GridView>标签显示商品类别，android:columnWidth="80dp"属性设置列的宽度为 80dp，android:numColumns="4" 属性设置网格视图为 4 列，android:stretchMode="spacingWidthUniform"属性设置均匀拉伸单元格的间距。

6．主视图的 Activity

```
/java/com/vt/c0403/MainActivity.java
15  public class MainActivity extends AppCompatActivity {
16      private Context mContext;
17      private List<CategoryModel> mCategoryModel;
18      private CategoryAdapter mAdapter;
19      private GridView mGridView;
20      @Override
21      protected void onCreate(Bundle savedInstanceState) {
22          super.onCreate(savedInstanceState);
23          setContentView(R.layout.activity_main);
24          mContext = MainActivity.this;
25          mCategoryModel = new ArrayList<>();
26          mCategoryModel.add(new CategoryModel(R.mipmap.img_category1, "图书音像"));
```

```
27        mCategoryModel.add(new CategoryModel(R.mipmap.img_category2, "日化护肤"));
28        mCategoryModel.add(new CategoryModel(R.mipmap.img_category3, "美食外卖"));
29        mCategoryModel.add(new CategoryModel(R.mipmap.img_category4, "鞋靴箱包"));
30        mCategoryModel.add(new CategoryModel(R.mipmap.img_category5, "食品饮料"));
31        mCategoryModel.add(new CategoryModel(R.mipmap.img_category6, "运动户外"));
32        mCategoryModel.add(new CategoryModel(R.mipmap.img_category7, "手机数码"));
33        mCategoryModel.add(new CategoryModel(R.mipmap.img_category8, "日用百货"));
34        mAdapter = new CategoryAdapter(mContext,mCategoryModel);
35        mGridView = findViewById(R.id.grid_view);
36        mGridView.setAdapter(mAdapter);//设置适配器
37        mGridView.setSelector(new ColorDrawable(Color.TRANSPARENT));//设置被选网格项的背景色
38        mGridView.setOnItemClickListener(new AdapterView.OnItemClickListener() {
39            @Override
40            public void onItemClick(AdapterView<?> parent, View view, int position, long id) {
41                Toast.makeText(mContext, "单击了第" + position + "项", Toast.LENGTH_SHORT).show();
42            }
43        });
44    }
45 }
```

第 26~33 行实例化 List< CategoryModel >对象，添加 8 个 CategoryModel 实例保存商品类别的图像资源 id 和商品类别的名称。第 34 行将上下文和保存 CategoryModel 实例的 List<CategoryModel>对象作为参数，实例化自定义的适配器——CategoryAdapter。第 35~43 行 GridView 对象获取布局文件中的 GridView 控件，然后设置适配器和被选网格项的背景色为透明色，最后添加单击网格项的监听事件。

7．测试运行

运行程序后，可以在"Logcat"窗口中观察输出的日志信息（如图 4-6 所示）。在 GridView 控件中，所有的网格项都同时显示出来，因此需要初始化 8 个 convertView 对象。

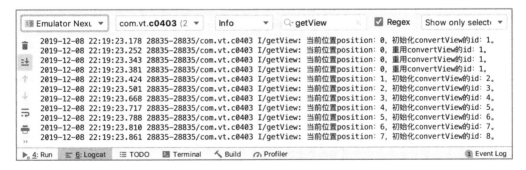

图 4-6 "Logcat"窗口输出的日志信息

4.4 自动完成文本视图

4.4.1 AutoCompleteTextView 控件

AutoCompleteTextView 控件（android.widget.AutoCompleteTextView）是根据输入文本

自动填充提示内容的输入和编辑文本的 UI 控件，是 android.widget.EditText 的子类，通过适配器将可供选择的提示文本数据与其匹配，除继承的标签属性外，还有 10 个标签属性（如表 4-5 所示），而类方法提供设置适配器、添加监听器、显示可选文本等更多功能（如表 4-6 所示）。Android 还提供了允许输入多个提示项的 MultiAutoCompleteTextView 控件，该控件的属性和方法与 AutoCompleteTextView 控件的属性和方法重合度很高。

表 4-5　AutoCompleteTextView 控件的 XML 标签属性

属　性	说　明
android:completionHint	设置下拉菜单中的提示标题
android:completionHintView	设置定义提示视图中显示的下拉菜单
android:completionThreshold	设置指定用户至少输入多少个字符才会显示提示
android:dropDownAnchor	设置下拉菜单的定位"锚点"组件。如果没有指定该属性，则使用该 TextView 作为定位"锚点"组件
android:dropDownHeight	设置下拉菜单的高度
android:dropDownWidth	设置下拉菜单的宽度
android:dropDownHorizontalOffset	设置下拉菜单与文本之间的水平间距
android:dropDownVerticalOffset	设置下拉菜单与文本之间的竖直间距
android:dropDownSelector	设置下拉菜单单击效果
android:popupBackground	设置下拉菜单的背景

表 4-6　AutoCompleteTextView 类的常用方法

类型和修饰符	方　法
boolean	isPopupShowing() 判断弹出菜单是否显示
int	getListSelection() 获取选择的列表项位置
<T extends ListAdapter & Filterable> void	setAdapter(T adapter) 设置适配器
void	setListSelection(int position) 设置选择的列表项
void	setOnClickListener(View.OnClickListener listener) 设置单击事件的监听器
void	setOnItemClickListener(AdapterView.OnItemClickListener l) 设置单击下拉表项的监听器
void	setOnItemSelectedListener(AdapterView.OnItemSelectedListener l) 设置下拉列表中选择列表项的监听器
void	showDropDown() 显示可选提示文本列表

4.4.2　实例工程：显示搜索提示的文本框

本实例演示了自动完成文本视图获取焦点后，显示相应的可选提示文本列表，以及多提示项自动完成文本视图可以多次根据输入的文本，显示相应的可选提示文本列表（如图 4-7 所示）。

图 4-7 运行效果

1．新建工程

新建一个"Empty Activity"工程，工程名称为"C0404"。

2．主界面的布局

```
/res/layout/activity_main.xml
01  <?xml version="1.0" encoding="utf-8"?>
02  <LinearLayout xmlns:android="http://schemas.android.com/apk/res/android"
03      android:layout_width="match_parent"
04      android:layout_height="match_parent"
05      android:orientation="horizontal">
06      <AutoCompleteTextView
07          android:id="@+id/actv"
08          android:layout_width="wrap_content"
09          android:layout_height="48dp"
10          android:layout_weight="1"
11          android:completionHint="请选择搜索内容"
12          android:completionThreshold="2"
13          android:hint="自动完成文本视图" />
14      <MultiAutoCompleteTextView
15          android:id="@+id/Mactv"
16          android:layout_width="wrap_content"
17          android:layout_height="48dp"
18          android:layout_weight="1"
19          android:completionThreshold="1"
20          android:dropDownHeight="200dp"
21          android:hint="多提示项自动完成文本视图" />
22  </LinearLayout>
```

第 06～13 行是<AutoCompleteTextView>标签，android:completionThreshold="2"表示输

入 2 个文本后显示可选提示文本列表。第 14～21 行是<MultiAutoCompleteTextView>标签，android:completionThreshold="1"表示输入 1 个文本后显示可选提示文本列表。

3. 主界面的 Activity

```
/java/com/vt/c0404/MainActivity.java
10   public class MainActivity extends AppCompatActivity {
11       private AutoCompleteTextView mActv;
12       private MultiAutoCompleteTextView mMactv;
13       private static final String[] data = new String[]{"蚊香", "香水", "香料", "香精", "香水品牌", "香水排名", "香水保质期", "女士香水"};
14       @Override
15       protected void onCreate(Bundle savedInstanceState) {
16           super.onCreate(savedInstanceState);
17           setContentView(R.layout.activity_main);
18           mActv = findViewById(R.id.actv);
19           mMactv = findViewById(R.id.Mactv);
20           //自动完成文本框适配数据
21           ArrayAdapter<String> arrayAdapter1 = new ArrayAdapter<>(MainActivity.this, android.R.layout.simple_dropdown_item_1line, data);
22           mActv.setAdapter(arrayAdapter1);
23           mActv.setOnFocusChangeListener(new View.OnFocusChangeListener() {
24               @Override
25               public void onFocusChange(View v, boolean hasFocus) {
26                   AutoCompleteTextView view = (AutoCompleteTextView) v;
27                   if (hasFocus) {
28                       view.showDropDown();//显示可选提示文本列表
29                   }
30               }
31           });
32           //多提示项自动完成文本框适配数据
33           ArrayAdapter<String> arrayAdapter2 = new ArrayAdapter<>(getApplicationContext(), android.R.layout.simple_dropdown_item_1line, data);
34           mMactv.setAdapter(arrayAdapter2);
35           mMactv.setTokenizer(new MultiAutoCompleteTextView.CommaTokenizer());//设置分隔符
36       }
37   }
```

第 13 行是可选提示文本的数组。第 21～31 行创建字符串数组适配器 arrayAdapter1 作为 mActv 的适配器，并添加焦点改变监听器，获取焦点时显示可选提示文本列表。第 33～35 行创建字符串数组适配器 arrayAdapter2 作为 mMactv 的适配器，并设置多项提示的分隔符。

4.5 悬 浮 框

4.5.1 PopupWindow 控件

PopupWindow 控件（android.widget.PopupWindow）是悬浮显示的 UI 控件，是 java.lang.Object 的子类，除继承的标签属性外，还有 6 个标签属性（如表 4-7 所示），而类方法还提供了根据位置显示悬浮框、关闭悬浮框等更多功能（如表 4-8 所示）。

表 4-7 PopupWindow 控件的 XML 标签属性

属性	说明
android:overlapAnchor	设置悬浮框是否应重叠其锚点视图
android:popupAnimationStyle	设置悬浮框的动画样式
android:popupBackground	设置悬浮框的背景
android:popupElevation	设置悬浮框的高度（z 轴的高度），默认值为 0dp
android:popupEnterTransition	设置悬浮框的显示过渡动画
android:popupExitTransition	设置悬浮框的消失过渡动画

表 4-8 PopupWindow 类的常用方法

类型和修饰符	方法
void	dismiss() 关闭悬浮框的视图
View	getContentView() 获取悬浮框的视图
void	setContentView(View contentView) 设置悬浮框的视图
void	showAsDropDown(View anchor) 在锚点视图左下角弹出悬浮框的视图
void	showAsDropDown(View anchor, int xoff, int yoff) 根据偏移值（像素）在锚点视图左下角弹出悬浮框的视图
void	showAtLocation(View parent, int gravity, int x, int y) 在指定视图位置（像素）弹出悬浮框的视图

4.5.2 实例工程：单击按钮显示自定义悬浮框

本实例演示了单击按钮显示悬浮框，单击悬浮框中的按钮显示提示信息且悬浮框消失（如图 4-8 所示），悬浮框显示和消失的过程使用动画效果。当悬浮框显示时，单击悬浮框以外的区域悬浮框会消失，且不会触发单击区域控件的事件。

图 4-8 运行效果

1. 打开基础工程

打开"基础工程"文件夹中的"C0405"工程,该工程已经包含图像资源、MainActivity及布局的文件。

2. 悬浮框背景图像

双击打开"/res/mipmap"资源文件夹中的"img_popup_bg.9.png"文件,在"9-Patch"图界面的图像四周,按住 Ctrl 键拖曳出黑色线条,按住 Shift 键拖曳可以删除多余的黑色线条区域。上方和左边的黑线分别表示横向和纵向可拉伸的区域,下方和右边的黑线分别表示横向和纵向显示内容的区域。勾选"Show content"和"Show patches"复选框,中间区域是可拉伸区域,右侧预览图显示了三种拉伸效果(如图 4-9 所示)。

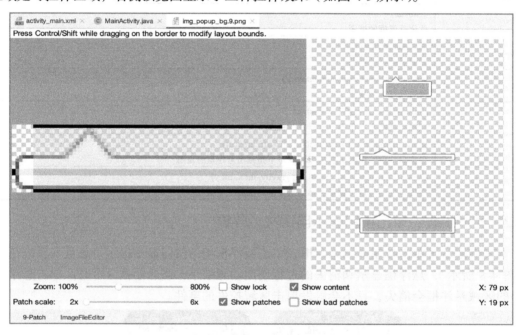

图 4-9　设置 9-Patch 图的可拉伸区域

 提示:9-Patch 图

　　9-Patch 图是一种可拉伸的位图,会自动调整大小使图像在充当背景时在界面中自适应。9-Patch 图是标准 PNG 格式的图像,并且将四周 1 像素宽作为边界的设置区域,设置成黑色表示可以拉伸的区域,程序运行时不会被显示出来。保存时要以"9.png"为扩展名,放置在项目的"/res/drawable"文件夹中。

3. 悬浮框的视图

在"/res/layout"文件夹中,新建"popup_share.xml"文件,用于创建弹出的悬浮框视图。

```
/res/layout/popup_share.xml
01  <?xml version="1.0" encoding="utf-8"?>
02  <LinearLayout xmlns:android="http://schemas.android.com/apk/res/android"
03      android:layout_width="match_parent"
```

```
04          android:layout_height="match_parent"
05          android:background="@mipmap/img_popup_bg"
06          android:orientation="vertical">
07          <Button
08              android:id="@+id/btn_weixin"
09              android:layout_width="wrap_content"
10              android:layout_height="wrap_content"
11              android:padding="5dp"
12              android:text="微信"
13              android:textSize="18sp" />
14          <Button
15              android:id="@+id/btn_weibo"
16              android:layout_width="wrap_content"
17              android:layout_height="wrap_content"
18              android:padding="5dp"
19              android:text="微博"
20              android:textSize="18sp" />
21      </LinearLayout>
```

第 05 行设置"img_popup_bg.png"图像作为 LinearLayout 控件的背景。第 07~20 行两个 Button 控件作为悬浮框的菜单按钮。

4．悬浮框出现的动画

选择"/res"文件夹，单击右键并选择【New】→【Android Resource Directory】命令（如图 4-10 所示）。在打开的"New Resource Directory"对话框中，设置"Directory name"为"anim"，"Resource type"为"anim"（如图 4-11 所示），单击"OK"按钮完成资源文件夹的创建。

图 4-10 【Android Resource Directory】命令

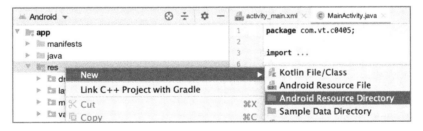

图 4-11 "New Resource Directory"对话框

选择"/res/anim"文件夹,单击右键并选择【New】→【Animation Resource File】命令(如图4-12所示)。在打开的"New Resource File"对话框中,设置"File name"为"popup_enter"(如图4-13所示),单击"OK"按钮新建"popup_enter.xml"文件,用于添加悬浮框出现的动画效果。

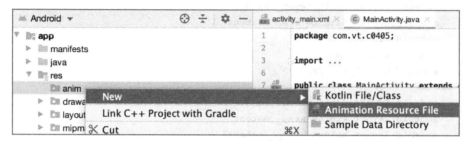

图 4-12 【Animation Resource File】命令

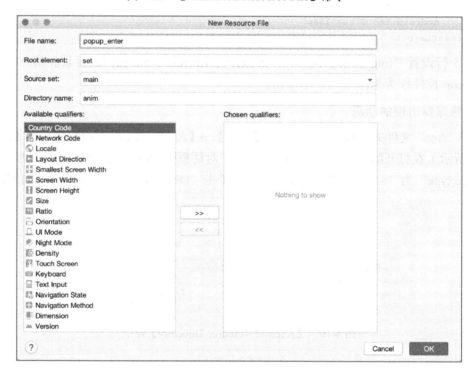

图 4-13 "New Resource File"对话框

```
/res/anim/popup_enter.xml
01  <?xml version="1.0" encoding="utf-8"?>
02  <set xmlns:android="http://schemas.android.com/apk/res/android">
03      <scale
04          android:duration="100"
05          android:fromXScale="0.6"
06          android:fromYScale="0.6"
07          android:pivotX="50%"
08          android:pivotY="50%"
09          android:toXScale="1.0"
```

```
10          android:toYScale="1.0" />
11      <translate
12          android:duration="200"
13          android:fromYDelta="10%"
14          android:toYDelta="0" />
15      <alpha
16          android:duration="100"
17          android:fromAlpha="0.0"
18          android:interpolator="@android:anim/decelerate_interpolator"
19          android:toAlpha="1.0" />
20  </set>
```

第 03～10 行<scale>标签设置缩放动画的参数,pivotX 属性和 pivotY 属性设置缩放中心点的 x、y 轴位置，属性值可以使用整数值、百分数（或小数）或百分数 p 的形式，如 50（表示相对于控件左上角的像素）、50%/0.5（表示相对于控件左上角的百分比）、50%p（表示对于父控件左上角的百分比）。第 11～14 行<translate>标签设置位移动画的参数，fromYDelta 属性和 toYDelta 属性设置 y 轴位移动画的起始位置，也可以使用整数值、百分数（或小数）或百分数 p 的形式。第 15～19 行<alpha>标签设置透明度动画的参数，interpolator 属性设置动画补间的差值器，@android:anim/decelerate_interpolator 是系统自带的减速差值器。

5．悬浮框消失的动画

在 "/res/anim" 文件夹中，新建 "popup_exit.xml" 动画资源文件，用于添加悬浮框消失的动画效果。

```
/res/anim/popup_exit.xml
01  <?xml version="1.0" encoding="utf-8"?>
02  <set xmlns:android="http://schemas.android.com/apk/res/android">
03      <scale
04          android:duration="500"
05          android:fromXScale="1.0"
06          android:fromYScale="1.0"
07          android:pivotX="50%"
08          android:pivotY="50%"
09          android:toXScale="0.5"
10          android:toYScale="0.5" />
11      <translate
12          android:duration="200"
13          android:fromYDelta="0"
14          android:toYDelta="10%" />
15      <alpha
16          android:duration="500"
17          android:fromAlpha="1.0"
18          android:interpolator="@android:anim/accelerate_interpolator"
19          android:toAlpha="0.0" />
20  </set>
```

第 03～10 行<scale>标签设置缩放动画的参数。第 11～14 行<translate>标签设置位移动画的参数。第 15～19 行<alpha>标签设置透明度动画的参数。

6. 悬浮框的动画

/res/values/styles.xml
10 `<style name="PopupAnimation" parent="android:Animation">`
11   `<item name="android:windowEnterAnimation">@anim/popup_enter</item>`
12   `<item name="android:windowExitAnimation">@anim/popup_exit</item>`
13 `</style>`

第 10 行<style>标签设置悬浮框的动画效果文件，name 属性表示动画效果的名称，使用 R.style.PopupAnimation 调用，parent 属性表示该样式所继承的样式。第 11、12 行两个<item>标签分别设置出现和消失的动画资源。

7. 主界面的 Activity

/java/com/vt/c0405/MainActivity.java
13 `public class MainActivity extends AppCompatActivity {`
14   `private Context mContext;`
15   `@Override`
16   `protected void onCreate(Bundle savedInstanceState) {`
17   `super.onCreate(savedInstanceState);`
18   `setContentView(R.layout.activity_main);`
19   `mContext = MainActivity.this;`
20   `Button replyBtn = findViewById(R.id.btn_reply);`
21   `replyBtn.setOnClickListener(new View.OnClickListener() {`
22   `@Override`
23   `public void onClick(View v) {`
24   `Toast.makeText(mContext, "回复信息", Toast.LENGTH_SHORT).show();`
25   `}`
26   `});`
27   `Button shareBtn = findViewById(R.id.btn_share);`
28   `shareBtn.setOnClickListener(new View.OnClickListener() {`
29   `@Override`
30   `public void onClick(View v) {`
31   `popWindow(v);`
32   `}`
33   `});`
34   `}`
35   `//初始化`
36   `private void popWindow(View v) {`
37   `//实例化悬浮框视图`
38   `View popupView = LayoutInflater.from(mContext).inflate(R.layout.popup_share, null, false);`
39   `//实例化悬浮框`
40   `final PopupWindow popupWindow = new PopupWindow(popupView, ViewGroup.LayoutParams.WRAP_CONTENT, ViewGroup.LayoutParams.WRAP_CONTENT, true);`
41   `popupWindow.setAnimationStyle(R.style.PopupAnimation);//设置动画`
42   `popupWindow.showAsDropDown(v, 10, 0);//显示悬浮框`
43   `//设置悬浮框视图中的按钮事件`
44   `Button weixinBtn = popupView.findViewById(R.id.btn_weixin);`
45   `Button weiboBtn = popupView.findViewById(R.id.btn_weibo);`
46   `weixinBtn.setOnClickListener(new View.OnClickListener() {`

```
47        @Override
48        public void onClick(View v) {
49            Toast.makeText(mContext, "已经分享到微信", Toast.LENGTH_SHORT).show();
50            popupWindow.dismiss();//关闭悬浮框
51        }
52    });
53    weiboBtn.setOnClickListener(new View.OnClickListener() {
54        @Override
55        public void onClick(View v) {
56            Toast.makeText(mContext, "已经分享到微信", Toast.LENGTH_SHORT).show();
57            popupWindow.dismiss();//关闭悬浮框
58        }
59    });
60  }
61 }
```

第 28～33 行 shareBtn 的单击监听事件中调用 intPopWindow(View v)方法，并将 onClick(View v)中的 v 参数作为参数继续传递下去。第 36～60 行设置并显示悬浮框，还处理了悬浮框中按钮的单击事件。第 40 行 PopupWindow 构造方法的第三个参数设置是否可以获取焦点，设置为 true 时，单击悬浮框以外的区域悬浮框会消失，不会触发单击区域的控件。

4.6 翻 转 视 图

4.6.1 ViewFlipper 控件

ViewFlipper 控件（android.widget.ViewFlipper）是多视图轮转的 UI 控件，是 android.widget.ViewAnimator 的子类。ViewFlipper 控件的常用 XML 标签属性如表 4-9 所示，ViewFlipper 类的常用方法如表 4-10 所示。

表 4-9 ViewFlipper 控件的常用 XML 标签属性

属　　性	说　　明
android:autoStart	设置是否自动开始翻转子视图
android:flipInterval	设置翻转子视图的间隔时间
android:inAnimation	设置子视图显示的动画
android:outAnimation	设置子视图消失的动画

表 4-10 ViewFlipper 类的常用方法

类型和修饰符	方　　法
int	getFlipInterval() 获取翻转子视图的间隔时间
boolean	isAutoStart() 判断是否自动开始翻转子视图
boolean	isFlipping() 判断是否正在翻转子视图

续表

类型和修饰符	方法
void	startFlipping() 开始翻转子视图
void	stopFlipping() 停止翻转子视图
void	addView(View child) 添加翻转子视图
void	addView(View child, int index) 添加翻转子视图到指定的索引号位置，子视图的索引号从 0 开始
View	getCurrentView() 获取当前翻转子视图
int	getDisplayedChild() 获取当前翻转子视图的索引号，子视图的索引号从 0 开始
void	removeAllViews() 移除所有翻转子视图
void	removeViewAt(int index) 移除指定索引号的翻转子视图，子视图的索引号从 0 开始
void	setOnClickListener(View.OnClickListener l) 设置单击时间监听器
void	showNext() 显示下一个翻转子视图
void	showPrevious() 显示上一个翻转子视图

4.6.2　实例工程：轮流显示图像的翻转视图

本实例演示了使用翻转视图轮转显示图像，单击图像后通过提示信息显示当前子视图的索引号（如图 4-14 所示）。由于翻转视图设置了显示动画，所以当子视图开始执行显示动画时当前子视图的索引号会改变。

1. 打开基础工程

打开"基础工程"文件夹中的"C0406"工程，该工程已经包含图像资源、MainActivity 及布局的文件。

2. 图像滚动出现的动画

在"/res"文件夹中，新建"anim"文件夹。然后在"/res/anim"文件夹中，新建"view_flipper_right_in.xml"文件，设置子视图显示的动画效果。

图 4-14　运行效果

```
/res/anim/view_flipper_right_in.xml
02  <set xmlns:android="http://schemas.android.com/apk/res/android">
03      <translate
04          android:duration="2000"
05          android:fromXDelta="100%p"
```

```
06          android:toXDelta="0" />
07    </set>
```

第 03～06 行<translate>标签设置动画持续时间为 2000 毫秒,沿着 x 轴从 0 坐标点处移动到距离 0 坐标点左侧父控件 100%宽度处。

3. 图像滚动离开的动画

在 "/res/anim" 文件夹中,新建 "view_flipper_left_out.xml" 文件,设置子视图消失的动画效果。

```
/res/anim/view_flipper_left_out.xml
02    <set xmlns:android="http://schemas.android.com/apk/res/android">
03        <translate
04            android:duration="2000"
05            android:fromXDelta="0"
06            android:toXDelta="-100%p" />
07    </set>
```

第 03～06 行<translate>标签设置动画持续时间为 2000 毫秒,沿着 x 轴从距离 0 坐标点左侧父控件 100%宽度处移动到 0 坐标点处。

4. 主界面的布局

```
/res/layout/activity_main.xml
10    <ViewFlipper
11        android:id="@+id/view_flipper"
12        android:layout_width="wrap_content"
13        android:layout_height="wrap_content"
14        android:layout_columnSpan="4"
15        android:flipInterval="3000"
16        android:inAnimation="@anim/view_flipper_right_in"
17        android:outAnimation="@anim/view_flipper_left_out">
18        <include layout="@layout/view_flipper_page1" />
19        <include layout="@layout/view_flipper_page2" />
20    </ViewFlipper>
```

第 15 行 ViewFlipper 控件设置翻转子视图的间隔时间。第 16～17 行设置子视图显示和消失的动画。第 18～19 行两个<include>标签设置子视图。

5. 主界面的 Activity

```
/java/com/vt/c0406/MainActivity.java
11    public class MainActivity extends AppCompatActivity {
12    private ViewFlipper mViewFlipper;
13        @Override
14        protected void onCreate(Bundle savedInstanceState) {
15            super.onCreate(savedInstanceState);
16            setContentView(R.layout.activity_main);
17            //通过布局文件获取子视图
18            View view3 = View.inflate(MainActivity.this, R.layout.view_flipper_page3, null);
```

```java
19          //创建图像视图作为子视图
20          ImageView imageView4 = new ImageView(this);
21          imageView4.setAdjustViewBounds(true);
22          imageView4.setImageResource(R.mipmap.img_9546);
23          //设置翻转视图
24          mViewFlipper = findViewById(R.id.view_flipper);
25          mViewFlipper.addView(imageView4);
26          mViewFlipper.addView(view3,2);
27          mViewFlipper.startFlipping();
28          mViewFlipper.setOnClickListener(new View.OnClickListener() {
29              @Override
30              public void onClick(View v) {
31                  Toast.makeText(MainActivity.this, "当前显示的子视图索引号为"+mViewFlipper.getDisplayedChild(), Toast.LENGTH_SHORT).show();
32              }
33          });
34          //开始按钮
35          Button startBtn = findViewById(R.id.button_start);
36          startBtn.setOnClickListener(new View.OnClickListener() {
37              @Override
38              public void onClick(View v) {
39                  mViewFlipper.startFlipping();
40              }
41          });
42          //停止按钮
43          Button stopBtn = findViewById(R.id.button_stop);
44          stopBtn.setOnClickListener(new View.OnClickListener() {
45              @Override
46              public void onClick(View v) {
47                  mViewFlipper.stopFlipping();
48              }
49          });
50          //上一页按钮
51          Button previousBtn = findViewById(R.id.button_previous);
52          previousBtn.setOnClickListener(new View.OnClickListener() {
53              @Override
54              public void onClick(View v) {
55                  mViewFlipper.showPrevious();
56              }
57          });
58          //下一页按钮
59          Button nextBtn = findViewById(R.id.button_next);
60          nextBtn.setOnClickListener(new View.OnClickListener() {
61              @Override
62              public void onClick(View v) {
63                  mViewFlipper.showNext();
64              }
65          });
66      }
67  }
```

第 25 行 mViewFlipper 添加 imageView4 作为子视图放在其他子视图的后面，mViewFlipper 已有两个子视图，因此 imageView4 作为第三个子视图的索引号为 2。第 26 行 mViewFlipper 添加 view3 作为子视图放在 2 号索引位置处，imageView4 的索引号顺位后移变为 3。第 31 行 mViewFlipper.getDisplayedChild()获取当前子视图的索引号，由于 mViewFlipper 通过标签设置了 android:inAnimation 属性，所以开始执行显示动画时当前子视图的索引号就变为该子视图的索引号。

4.7 分 页 视 图

4.7.1 ViewPager 控件

ViewPager 控件（androidx.viewpager.widget.ViewPager）是分页滚动的 UI 控件，是 android.view.ViewGroup 的子类。该控件没有独有的标签属性，其标签属性都继承自父类。ViewPager 类的常用方法如表 4-11 所示。

表 4-11 ViewPager 类的常用方法

类型和修饰符	方 法
void	addOnAdapterChangeListener(ViewPager.OnAdapterChangeListener listener) 添加更改适配器的监听器
void	addOnPageChangeListener(ViewPager.OnPageChangeListener listener) 添加页面改变的监听器
PagerAdapter	getAdapter() 获取适配器
int	getCurrentItem() 获取当前分页的索引号
void	setAdapter(PagerAdapter adapter) 设置适配器
void	setCurrentItem(int item) 跳转到指定索引号的分页
void	setCurrentItem(int item, boolean smoothScroll) 是否平滑滚动到指定索引号的分页

PagerAdapter 类为 ViewPager 控件提供数据适配的功能（如表 4-12 所示），PagerAdapter 子类需要重写 4 个方法：instantiateItem(ViewGroup container, int position)、destroyItem(ViewGroup container, int position, Object object)、getCount()和 isViewFromObject(View view, Object object)。

表 4-12 PagerAdapter 类的常用方法

类型和修饰符	方 法
Object	instantiateItem(ViewGroup container, int position) 为指定索引位置创建分页视图。适配器负责将视图添加到此处给定的容器中，从 finishUpdate(ViewGroup container）返回时调用该方法
void	destroyItem(ViewGroup container, int position, Object object) 删除指定索引位置的分页视图

续表

类型和修饰符	方法
void	finishUpdate(ViewGroup container) 当前分页视图更新完成时调用
int	getCount() 获取分页视图的数量
boolean	isViewFromObject(View view, Object object) 判断视图是否与对象相关联
void	notifyDataSetChanged() 数据改变时更新视图
void	startUpdate(ViewGroup container) 显示的分页视图更新开始时调用此方法

ViewPager.OnPageChangeListener 接口用于监听 ViewPager 控件的事件，包含 3 个抽象方法（如表 4-13 所示）。

表 4-13 ViewPager.OnPageChangeListener 接口的方法

类型和修饰符	方法
abstract void	onPageScrollStateChanged(int state) 页面滚动状态改变事件，state 参数是事件状态，包括 SCROLL_STATE_IDLE、SCROLL_STATE_DRAGGING、SCROLL_STATE_SETTLING
abstract void	onPageScrolled(int position, float positionOffset, int positionOffsetPixels) 页面滚动事件，position 参数是分页视图的索引号，positionOffset 参数是分页视图的偏移百分比，positionOffsetPixels 参数是分页视图的偏移像素
abstract void	onPageSelected(int position) 页面选择事件，position 参数是分页视图的索引号

4.7.2 实例工程：欢迎引导页

本实例演示了首次使用 App 时显示的欢迎引导页，该页面包含 4 个分页，左右滑动、单击左右箭头或底部 4 个圆点可以切换分页。切换分页时，底部的 4 个圆点标识显示的分页位置（如图 4-15 所示）。为了便于演示监听事件，添加了两个 TextView 控件放置于顶部和底部，用于直观显示监听信息。

1. 打开基础工程

打开"基础工程"文件夹中的"C0407"工程，该工程已经包含图像资源、MainActivity 及布局的文件。

2. 主题的样式

图 4-15 运行效果

```
/res/values/styles.xml
03    <style name="AppTheme" parent="Theme.AppCompat.Light.NoActionBar">
04        <item name="android:windowBackground">@mipmap/img_loding_bg</item>
05        <item name="colorPrimary">@color/colorPrimary</item>
```

```
06        <item name="colorPrimaryDark">@color/colorPrimaryDark</item>
07        <item name="colorAccent">@color/colorAccent</item>
08    </style>
```

第 03 行将 Theme.AppCompat.Light.DarkActionBar 修改为 Theme.AppCompat.Light.NoActionBar，取消显示页面顶部的 ActionBar。第 04 行添加 android:windowBackground 属性，为所有的 Activity 设置背景图像，主要作用是避免在启动 App 的加载期间显示空白背景，增强用户体验。第 05 行设置主题的 tabar 背景颜色。第 06 行设置主题的状态栏背景颜色。第 07 行设置控件重点外观元素的颜色，如单选按钮的圆点颜色、复选框的方块颜色、对话框的默认按钮文字颜色。

 提示：主题

主题（theme）是 App 界面元素显示效果的规则，Android Studio 3.6 创建的 Empty Activity 项目使用的默认主题继承自 Theme.AppCompat.Light.DarkActionBar。主题通过"AndroidManifest.xml"文件中<application>标签的 android:theme 属性设置。

3. 状态栏的背景颜色

```
/res/values/colors.xml
02    <resources>
03        <color name="colorPrimary">#008577</color>
04        <color name="colorPrimaryDark">#dd4661</color>
05        <color name="colorAccent">#D81B60</color>
06    </resources>
```

第 04 行修改<color>标签的 colorPrimaryDark 属性值为#dd4661，该颜色与背景图像 @mipmap/img_loding_bg 的背景颜色相同，消除了状态栏与页面之间的分割感。

4. 分页视图的适配器

在"/java/com/vt/c0407"文件夹中，新建"WelcomePagerAdapter.java"文件。WelcomePagerAdapter 类继承自 PagerAdapter 类，用于自定义分页视图的适配器。

```
/java/com/vt/c0407/WelcomePagerAdapter.java
08    public class WelcomePagerAdapter extends PagerAdapter {
09        private ArrayList<View> mPagerViews;   //分页视图的数组列表
10        //构造方法
11        WelcomePagerAdapter(ArrayList<View> pagerViews) {
12            mPagerViews = pagerViews;
13        }
14        //获取分页视图的数量
15        @Override
16        public int getCount() {
17            return mPagerViews.size();
18        }
19        //判断是否由对象生成界面
20        @Override
21        public boolean isViewFromObject(View view, Object object) {
22            return view == object;
```

```
23        }
24    //显示分页视图或缓存分页时进行布局的初始化
25    @Override
26    public Object instantiateItem(ViewGroup container, int position) {
27        container.addView(mPagerViews.get(position));
28        return mPagerViews.get(position);
29    }
30    //销毁分页视图时移除相应的分页
31    @Override
32    public void destroyItem(ViewGroup container, int position, Object object) {
33        container.removeView(mPagerViews.get(position));
34    }
35 }
```

第 11 行将分页视图的数组列表作为构造方法的参数传递进来。第 22 行使用官方建议的返回值表达式 view==object，如果直接使用 true 或 false 作为返回值，会出现显示错误。第 27 行在初始化或缓存分页视图时将相应的分页视图添加到使用该适配器的控件中，ViewPager 控件作为 ViewGroup 的子类可以自动向上转为 ViewGroup 类实例。第 28 行将初始化或缓存分页视图作为返回对象。第 33 行在销毁分页面时将该分页面从 container 对象中移除。

5．主界面的布局

```
/res/layout/activity_main.xml
07    <!-- 分页视图 -->
08    <androidx.viewpager.widget.ViewPager
09        android:id="@+id/view_pager"
10        android:layout_width="match_parent"
11        android:layout_height="match_parent"
12        android:layout_gravity="center"
13        android:persistentDrawingCache="animation">
14    </androidx.viewpager.widget.ViewPager>
15    <!-- 显示滚动事件信息 -->
16    <TextView
17        android:id="@+id/text_view_page_position"
18        android:layout_width="wrap_content"
19        android:layout_height="wrap_content"
20        android:layout_gravity="top|center_horizontal"
21        android:layout_marginTop="25dp"
22        android:textColor="#ffffff" />
23    <!-- 前进按钮 -->
24    <ImageView
25        android:id="@+id/image_button_left"
26        android:layout_width="wrap_content"
27        android:layout_height="wrap_content"
28        android:layout_gravity="bottom|left"
29        android:layout_marginStart="10dp"
30        android:layout_marginBottom="60dp"
31        android:src="@mipmap/img_left" />
32    <!-- 后退按钮 -->
```

33	`    <ImageView`
34	`        android:id="@+id/image_button_right"`
35	`        android:layout_width="wrap_content"`
36	`        android:layout_height="wrap_content"`
37	`        android:layout_gravity="bottom|right"`
38	`        android:layout_marginEnd="10dp"`
39	`        android:layout_marginBottom="60dp"`
40	`        android:src="@mipmap/img_right" />`
41	`    <!-- 分页标记提示 -->`
42	`    <LinearLayout`
43	`        android:layout_width="match_parent"`
44	`        android:layout_height="wrap_content"`
45	`        android:layout_gravity="bottom"`
46	`        android:layout_marginBottom="65dp"`
47	`        android:gravity="center">`
48	`        <ImageView`
49	`            android:id="@+id/image_view_page0"`
50	`            android:layout_width="10dp"`
51	`            android:layout_height="10dp"`
52	`            android:scaleType="fitXY"`
53	`            android:src="@mipmap/img_page_now" />`
54	`        <ImageView`
55	`            android:id="@+id/image_view_page1"`
56	`            android:layout_width="10dp"`
57	`            android:layout_height="10dp"`
58	`            android:layout_marginLeft="10dp"`
59	`            android:scaleType="fitXY"`
60	`            android:src="@mipmap/img_page" />`
61	`        <ImageView`
62	`            android:id="@+id/image_view_page2"`
63	`            android:layout_width="10dp"`
64	`            android:layout_height="10dp"`
65	`            android:layout_marginLeft="10dp"`
66	`            android:scaleType="fitXY"`
67	`            android:src="@mipmap/img_page" />`
68	`        <ImageView`
69	`            android:id="@+id/image_view_page3"`
70	`            android:layout_width="10dp"`
71	`            android:layout_height="10dp"`
72	`            android:layout_marginLeft="10dp"`
73	`            android:scaleType="fitXY"`
74	`            android:src="@mipmap/img_page" />`
75	`    </LinearLayout>`

第 08~14 行添加 ViewPager 控件，android:persistentDrawingCache="animation"表示在布局动画后进行缓存。第 20 行使用 "|" 连接两个属性值，top|center_horizontal 属性值表示顶部对齐且水平居中。第 42~75 行在<LinearLayout>标签内添加了 4 个<ImageView>标签，标识当前显示的分页。

6. 主界面的 Activity

/java/com/vt/c0407/MainActivity.java

```java
43   //初始化
44   private void init(){
45       ArrayList<View> pagerViews = new ArrayList<>();
46       pagerViews.add(View.inflate(this, R.layout.view_pager_welcome0, null));
47       pagerViews.add(View.inflate(this, R.layout.view_pager_welcome1, null));
48       pagerViews.add(View.inflate(this, R.layout.view_pager_welcome2, null));
49       pagerViews.add(View.inflate(this, R.layout.view_pager_welcome3, null));
50       WelcomePagerAdapter welcomePagerAdapter = new WelcomePagerAdapter(pagerViews);
51       mViewPager.setAdapter(welcomePagerAdapter);
52   }
53   //单击监听事件
54   @Override
55   public void onClick(View v) {
56       switch (v.getId()) {
57           case R.id.image_button_left:
58               if (mCurrentPosition > 0) { mCurrentPosition--; }
59               mViewPager.setCurrentItem(mCurrentPosition, true);
60               break;
61           case R.id.image_button_right:
62               if (mCurrentPosition < 3) { mCurrentPosition++; }
63               mViewPager.setCurrentItem(mCurrentPosition, true);
64               break;
65           case R.id.image_view_page0:
66               mCurrentPosition = 0;
67               mViewPager.setCurrentItem(mCurrentPosition, true);
68               break;
69           case R.id.image_view_page1:
70               mCurrentPosition = 1;
71               mViewPager.setCurrentItem(mCurrentPosition, true);
72               break;
73           case R.id.image_view_page2:
74               mCurrentPosition = 2;
75               mViewPager.setCurrentItem(mCurrentPosition, true);
76               break;
77           case R.id.image_view_page3:
78               mCurrentPosition = 3;
79               mViewPager.setCurrentItem(mCurrentPosition, true);
80               break;
81       }
82   }
83   //页面改变监听器
84   public class PageChangeListener implements ViewPager.OnPageChangeListener {
85       //页面选择事件
86       public void onPageSelected(int position) {
87           //翻页时当前page, 改变当前状态圆点图像
88           mCurrentPosition = position;
89           switch (position) {
90               case 0:
```

```
 91             mPage0.setImageResource(R.mipmap.img_page_now);
 92             mPage1.setImageResource(R.mipmap.img_page);
 93             mPage2.setImageResource(R.mipmap.img_page);
 94             mPage3.setImageResource(R.mipmap.img_page);
 95             break;
 96         case 1:
 97             mPage1.setImageResource(R.mipmap.img_page_now);
 98             mPage0.setImageResource(R.mipmap.img_page);
 99             mPage2.setImageResource(R.mipmap.img_page);
100             mPage3.setImageResource(R.mipmap.img_page);
101             break;
102         case 2:
103             mPage2.setImageResource(R.mipmap.img_page_now);
104             mPage0.setImageResource(R.mipmap.img_page);
105             mPage1.setImageResource(R.mipmap.img_page);
106             mPage3.setImageResource(R.mipmap.img_page);
107             break;
108         case 3:
109             mPage3.setImageResource(R.mipmap.img_page_now);
110             mPage0.setImageResource(R.mipmap.img_page);
111             mPage1.setImageResource(R.mipmap.img_page);
112             mPage2.setImageResource(R.mipmap.img_page);
113             break;
114         }
115     }
116     //页面滚动事件
117     @SuppressLint("SetTextI18n")
118     public void onPageScrolled(int position, float positionOffset, int positionOffsetPixels){
119         mPagePositionTextView.setText("页面索引号:" + position + "  偏移百分比:" + positionOffset + "  偏移像素:" + positionOffsetPixels);
120     }
121     //页面滚动状态改变事件
122     public void onPageScrollStateChanged(int state) {
123         switch (state) {
124         case ViewPager.SCROLL_STATE_IDLE:
125             mPageStateTextView.setText("空闲状态");
126             break;
127         case ViewPager.SCROLL_STATE_DRAGGING:
128             mPageStateTextView.setText("拖动状态");
129             break;
130         case ViewPager.SCROLL_STATE_SETTLING:
131             mPageStateTextView.setText("结束状态");
132             break;
133         }
134     }
135 }
```

第 43 行实例化一个 PageChangeListener 类对象作为监听器。第 44～52 行初始化并设置适配器。第 84～135 行的 PageChangeListener 类实现 3 个 ViewPager.OnPageChangeListener 接口方法。

4.8 习　　题

1．实现淘宝搜索后界面显示商品列表的效果，至少包含 8 个商品。

2．实现微信右上角⊕号的悬浮菜单效果，菜单包含发起群聊、添加朋友、扫一扫、收付款和帮助与反馈。

3．实现京东首页的轮播广告效果，至少包含 3 个产品广告的图像。

第 5 章 基本程序单元

UI 控件无法单独使用呈现给用户，无论使用标签形式的布局文件还是通过代码动态创建的 UI 控件，都需要通过基本程序单元作为载体进行呈现。Activity 和 Fragment 类似于页面的功能，其主要功能是将 UI 控件呈现给用户，并与用户进行交互。

5.1 活　　动

5.1.1 Activity 概述

Activity（android.app.Activity）是以窗口形式显示的组件（如表 5-1 所示），App 通常由多个彼此松散绑定的 Activity（活动）组成。

表 5-1　Activity 类的常用方法

类型和修饰符	方　　法
void	addContentView(View view, ViewGroup.LayoutParams params) 添加内容视图
void	finish() 关闭 Activity
void	finishActivity(int requestCode) 关闭使用 startActivityForResult（Intent，int）启动的 Activity
Intent	getIntent() 获取启动 Activity 的 Intent 对象
void	setContentView(int layoutResID) 设置 Activity 的布局资源
void	startActivity(Intent intent) 启动 Activity
void	startActivity(Intent intent, Bundle options) 启动 Activity 并传递数据
void	startActivityForResult(Intent intent, int requestCode) 启动 Activity 并传递请求码
void	startActivityForResult(Intent intent, int requestCode, Bundle options) 启动 Activity，在关闭该 Activity 时能够获取返回的数据
void	onNewIntent(Intent intent) 重新调用栈内 Activity 到前台时，调用该方法
void	onCreate(Bundle savedInstanceState) 创建时调用该方法

续表

类型和修饰符	方　　法
void	onRestart() 即将重新启动时调用该方法
void	onStart() 即将启动时调用该方法
void	onResume() 即将进入前台时调用该方法
void	onPause() 即将进入后台时调用该方法
void	onStop() 进入后台后调用该方法
void	onDestroy() 即将销毁时调用该方法
void	onActivityResult(int requestCode, int resultCode, Intent data) 当使用 startActivityForResult()方法启动的 Activity 退出时调用该方法，并将请求码、返回码及附加数据回传

Activity 需要在 AndroidManifest.xml 中添加<activity>标签后才能被调用，<activity>标签属性（如表 5-2 所示）大多无法通过 Activity 类的方法进行设置和获取。

表 5-2 <activity>标签属性

属　　性	说　　明
android:alwaysRetainTaskState	设置是否始终保持 Activity 所在任务的状态而不被系统回收，该属性只对任务的根 Activity 设置有效
android:directBootAware	是否支持直接启动，即是否可以在用户解锁设备前运行
android:enabled	是否可实例化 Activity
android:excludeFromRecents	是否从最近任务列表中删除该 Activity 启动的任务
android:exported	设置是否可由其他 App 的组件启动
android:icon	设置图标
android:immersive	设置是否使用沉浸模式
android:label	设置标签
android:launchMode	设置启动模式。可选值包括 standard、singleTop、singleTask 和 singleInstance
android:multiprocess	设置是否可将 Activity 实例启动到该实例的组件进程中，默认值为 false
android:name	设置实现 Activity 的类名
android:process	设置运行 Activity 的进程名称。如果名称以 "：" 开头，则系统在需要时会创建专用新进程运行 Activity；如果名称以小写字母开头，则 Activity 将在使用该名称的全局进程中运行
android:resizeableActivity	设置是否支持多窗口显示。设置为 true 时，可以在分屏和自由窗口模式下启动；设置为 false 时，在多窗口模式下启动将全屏显示
android:screenOrientation	设置非多窗口模式下屏幕显示的方向。 ● unspecified：默认值，系统决定方向 ● behind：与栈中其后的 Activity 方向相同 ● landscape：横向显示 ● portrait：纵向显示 ● reverseLandscape：与正常横向方向相反的横向 ● reversePortrait：与正常纵向方向相反的纵向 ● sensorLandscape：横向显示，根据传感器调整为正常或反向的横向 ● sensorPortrait：纵向显示，根据传感器调整为正常或反向的纵向

续表

属　性	说　明
android:screenOrientation	● userLandscape：横向显示，根据传感器和用户首选项调整为正常或反向的横向 ● userPortrait：纵向显示，根据传感器和用户首选项调整为正常或反向的纵向 ● sensor：屏幕方向由设备方向传感器决定，正常情况下某些设备不使用反向纵向或反向横向 ● fullSensor：屏幕方向由设备方向传感器决定，该值均支持所有 4 种可能的屏幕方向 ● nosensor：忽略传感器 ● user：用户当前的首选方向 ● fullUser：如果用户锁定基于传感器的旋转，则与 user 属性值相同，否则与 fullSensor 属性值相同 ● locked：锁定为其当前的方向
android:supportsPictureInPicture	设置是否支持画中画显示功能。如果 android:resizeableActivity 为 false，则系统会忽略该属性

 提示：AppCompatActivity

AppCompatActivity 类间接继承自 Activity 类，与 Activity 类相比，在视图顶部增加了一个 ActionBar，可以设置 Material 风格及直接使用 Material 风格，如 toolBar、Snackbar 和 AlertDialog 等。使用 AppCompatActivity 或直接使用 AppCompateDelegate，都必须使用 Theme.AppCompat 样式。

5.1.2　Activity 的创建和删除

1．创建 Activity

新建一个"Empty Activity"工程，工程名称为"C0501"。在"AndroidManifest.xml"文件中，已经注册了该 Activity 的配置信息，并在<intent-filter>标签中设置该 Activity 为默认启动项。

```
/Manifests/AndroidManifest.xml
10    <activity android:name=".MainActivity">
11        <intent-filter>
12            <action android:name="android.intent.action.MAIN" />
13            <category android:name="android.intent.category.LAUNCHER" />
14        </intent-filter>
15    </activity>
```

在 MainActivity 所在的文件夹上单击右键，选择【New】→【Activity】子菜单下的相应 Activity 预设可以快速创建 Activity（如图 5-1 所示）。

在打开的"New Android Activity"对话框中，默认勾选"Generate Layout File"复选框，用于同时创建一个布局文件（如图 5-2 所示）。"Launcher Activity"复选框用于设置工程的默认启动项，勾选后会替代之前的启动项作为默认启动的 Activity。

创建完成后，会在"AndroidManifest.xml"文件中自动添加<activity android:name=".InnerActivity"> </activity>代码，用于注册 Activity。

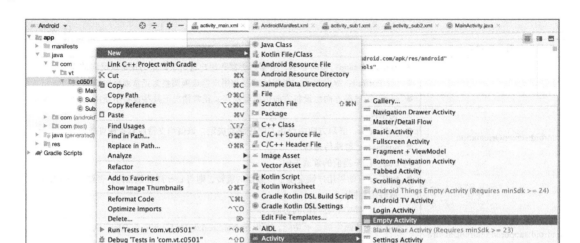

图 5-1 创建 Activity 的子菜单

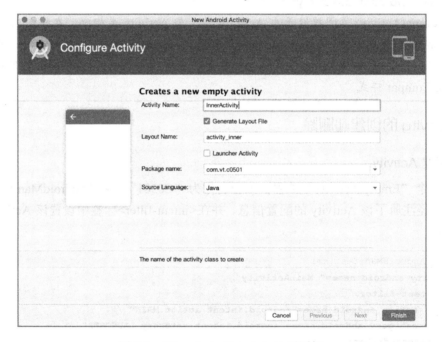

图 5-2 "New Android Activity" 对话框

2．删除 Activity

在 Project 窗口的 InnerActivity 上单击右键，选择【Delete】命令，打开"Delete"对话框（如图 5-3 所示），单击"OK"按钮进行删除。如果勾选所有复选框，会将与该 Activity 唯一相关的配置信息和文本资源都删除，否则需要手动删除相应选项的内容。

删除 Activity 后，并不会删除创建该 Activity 时的布局文件。在 Project 窗口的 layout 文件夹的 activity_inner.xml 文件上单击右键，选择【Delete】命令，打开"Delete"对话框（如图 5-4 所示），如果勾选两个复选框，会将与该布局文件唯一相关的内容删除掉，否则需要手动删除相应选项的内容。

图 5-3 "Delete" 对话框（1）　　　　图 5-4 "Delete" 对话框（2）

5.1.3　Activity 的启动和关闭

1. 启动 Activity

通过启动项的 Activity 启动其他 Activity 时，最简单的方法是使用 startActivity（Intent intent）方法，该方法需要使用一个 Intent 对象。Intent 是一个消息传递对象，用于从其他应用组件请求操作。Intent 可以通过多种方式促进组件之间的通信，主要功能是启动 Activity、启动 Service 和传递广播。

> **提示：显式 Intent 和隐式 Intent**
>
> 　　显式 Intent 直接通过组件名指定启动的组件，会明确指定启动的组件路径。
>
> 　　隐式 Intent 不直接指定组件名，而指定 Intent 的 Action、Data 或 Category，多用于调用系统默认提供的功能。当启动组件时，自动匹配系统中所有安装 App 的 AndroidManifest.xml 文件中的<intent-filter>标签，匹配出满足条件的组件。当不止一个组件满足条件时，会打开一个选择组件的对话框。

打开"基础工程"文件夹中的 C0502 工程，MainActivity 演示了 8 种打开 Activity 的方式，分为显式和隐式两种类型。其中，2 个以显式的方式启动，1 个以自定义隐式的方式启动，5 个使用预设隐式的方式启动。

- 启动内部 Activity（显式）：InnerActivity 是同一工程中的 Activity，后缀名使用 class。

```
Intent intent = new Intent(MainActivity.this, InnerActivity.class);
startActivity(intent);
```

- 启动外部 Activity（显式）：com.vt.c0407 是外部包名，com.vt.c0407.MainActivity 是外部包中的类名。

```
Intent intent = new Intent("android.intent.action.MAIN");
intent.setClassName("com.vt.c0407", "com.vt.c0407.MainActivity");
startActivity(intent);
```

- 启动外部 Activity（预设隐式）：MediaStore.INTENT_ACTION_STILL_IMAGE_CAMERA 设置打开拍摄照片的组件。如果没有安装第三方的拍摄照片 App，则直接调用系统相机 App；如果安装了第三方的拍摄照片 App，则需要进行选择。

```
Intent intent = new Intent();
intent.setAction(MediaStore.INTENT_ACTION_STILL_IMAGE_CAMERA);
startActivity(intent);
```

在C0501工程的AndroidManifest.xml文件中，设置Sub1Activity为拍照功能的Activity。安装C0501后，需要选择启动哪个组件（如图5-5所示）。

```
/manifests/AndroidManifest.xml
16    <activity android:name=".Sub1Activity">
17        <intent-filter>
18            <action android:name="android.media.action.STILL_IMAGE_CAMERA" />
19            <category android:name="android.intent.category.DEFAULT" />
20        </intent-filter>
21    </activity>
```

图 5-5　选择启动的组件

- 启动外部 Activity（自定义隐式）：intent.setAction（"custom_action"）设置外部Activity 的<action>标签名称，intent.addCategory（"custom_category"）设置外部Activity 的<category>标签名称，这两个标签属性相对应的 Activity 是在 C0501 工程中的 Sub2Activity。安装 C0501 后，才能启动 C0501 的 Sub2Activity。

```
Intent intent = new Intent();
intent.setAction("custom_action");
intent.addCategory("custom_category");
startActivity(intent);
```

在 C0501 工程的 AndroidManifest.xml 文件中，设置 Sub2Activity 的<action>和<category>标签。

```
/manifests/AndroidManifest.xml
22    <activity android:name=".Sub2Activity">
23        <intent-filter>
24            <action android:name="custom_action" />
25            <category android:name="custom_category" />
```

```
26          <category android:name="android.intent.category.DEFAULT" />
27      </intent-filter>
28  </activity>
```

- 启动拨打电话（预设隐式）：Intent.ACTION_DIAL 设置启动拨号组件，uri 设置拨打的电话号码。

```
Uri uri = Uri.parse("tel:10086");
Intent intent = new Intent(Intent.ACTION_DIAL, uri);
startActivity(intent);
```

- 启动发送短信（预设隐式）：Intent.ACTION_SENDTO 设置启动发送指定目标的组件，uri 设置接收短信的电话号码，putExtra("sms_body", "短信内容")设置发送短信的内容为"测试短信"。系统根据 sms_body 确定启动发送短信。

```
Uri uri = Uri.parse("smsto:10086");
Intent intent = new Intent(Intent.ACTION_SENDTO, uri);
intent.putExtra("sms_body", "测试短信");
startActivity(intent);
```

- 启动发送邮件（预设隐式）：Intent.ACTION_SEND 设置启动发送的组件，putExtra(Intent.EXTRA_EMAIL, "baizhe_22@qq.com")设置接收邮箱的地址，putExtra(Intent.EXTRA_SUBJECT,"邮件标题")设置邮件的标题，putExtra(Intent.EXTRA_TEXT,"邮件内容")设置邮件的内容，setType("text/plain")设置纯文本的邮件并未包含附件。

```
Intent intent = new Intent(Intent.ACTION_SEND);
intent.putExtra(Intent.EXTRA_EMAIL, "baizhe_22@qq.com");
intent.putExtra(Intent.EXTRA_SUBJECT, "邮件标题");
intent.putExtra(Intent.EXTRA_TEXT, "邮件内容");
intent.setType("text/plain");
startActivity(intent);
```

- 启动浏览器（预设隐式）：Intent.ACTION_VIEW 设置启动显示数据的组件，uri 设置访问的网址。

```
Uri uri = Uri.parse("http://www.weiju2014.com");
Intent intent = new Intent(Intent.ACTION_VIEW, uri);
startActivity(intent);
```

2. 关闭 Activity

在 C0502 工程的 InnerActivity 中，调用 Activity 的 finish()方法关闭 Activity。

```
InnerActivity.this.finish();
```

5.1.4 Activity 的生命周期

Activity 的生命周期（如图 5-6 所示）主要包含 7 个方法和 6 个状态（Created、Started、Resumed、Paused、Stopped、Destroyed）。完整生命周期从回调 onCreate()方法开始，到回

调 onDestroy()方法结束。可见生命周期从回调 onStart()方法开始，到回调 onStop()方法结束。前台生命周期从回调 onResume()方法开始，到回调 onPause()方法结束。

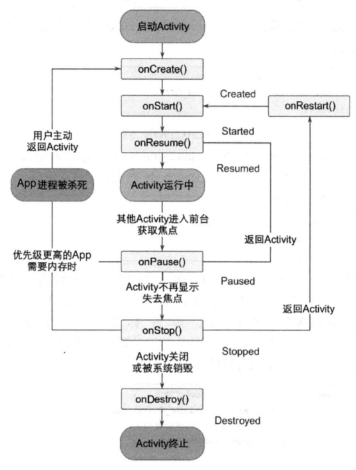

图 5-6　Activity 的生命周期

打开"基础工程"文件夹中的 C0503 工程，MainActivity 重写 7 个生命周期方法，用于测试 Activity 的生命周期执行顺序。SubActivity 主要将 Log.d（String tag, String msg）方法中的 MainActivity 替换为 SubActivity，onClick（View v）方法中的启动 SubActivity 替换为关闭 SubActivity。

```
/java/com/vt/c0503/MainActivity.java
11    final String TAG ="生命周期";
12    //Activity创建时调用
13    @Override
14    public void onCreate(Bundle savedInstanceState) {
15        super.onCreate(savedInstanceState);
16        setContentView(R.layout.activity_main);
17        Log.d(TAG, "MainActivity.onCreate()");
18        Button button= findViewById(R.id.button);
19        button.setOnClickListener(new View.OnClickListener() {
20            @Override
```

```
21        public void onClick(View v) {
22            Intent intent = new Intent(MainActivity.this, SubActivity.class);
23            startActivity(intent);
24            Log.d(TAG, "打开 SubActivity");
25        }
26    });
27 }
28 //Activity 即将启动时调用
29 @Override
30 protected void onStart() {
31     super.onStart();
32     Log.d(TAG, "MainActivity.onStart()");
33 }
34 //Activity 即将重新启动时调用
35 @Override
36 protected void onRestart() {
37     super.onRestart();
38     Log.d(TAG, "MainActivity.onRestart()");
39 }
40 //Activity 即将进入后台时调用
41 @Override
42 protected void onResume() {
43     super.onResume();
44     Log.d(TAG, "MainActivity.onResume()");
45 }
46 //Activity 即将进入后台时调用
47 @Override
48 protected void onPause() {
49     super.onPause();
50     Log.d(TAG, "MainActivity.onPause()");
51 }
52 //Activity 进入后台后调用
53 @Override
54 protected void onStop() {
55     super.onStop();
56     Log.d(TAG, "MainActivity.onStop()");
57 }
58 //Activity 即将被销毁时调用
59 @Override
60 public void onDestroy() {
61     super.onDestroy();
62     Log.d(TAG, "MainActivity.onDestroy()");
63 }
```

运行工程后，在"Logcat"窗口的过滤器中输入"生命周期"作为过滤标签，查看过滤后的结果，可以查看启动后执行的生命周期方法名称（如图 5-7 所示）。

在"Logcat"窗口中，单击右键并选择【Clear logcat】命令清空窗口。然后单击"打开 Activity"按钮启动 SubActivity，可以查看已经执行的生命周期方法名称（如图 5-8 所示）。

图 5-7 "Logcat"窗口输出的结果（1）

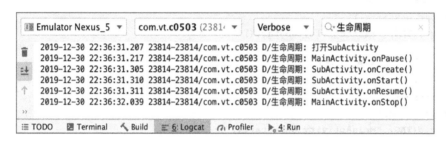

图 5-8 "Logcat"窗口输出的结果（2）

在"Logcat"窗口中，单击右键并选择【Clear logcat】命令清空窗口。然后单击"关闭 Activity"按钮或返回按钮关闭 SubActivity，可以查看已经执行的生命周期方法名称（如图 5-9 所示）。

图 5-9 "Logcat"窗口输出的结果（3）

在"Logcat"窗口中，单击右键并选择【Clear logcat】命令清空窗口。然后单击多任务按钮（底部正方形的虚拟按钮，如图 5-10 所示），快速单击 C0503 返回 MainActivity，可以查看已经执行的生命周期方法名称（如图 5-11 所示）。如果单击 C0503 的速度足够快，则不会调用 onRestart()方法。

在"Logcat"窗口中，单击右键并选择【Clear logcat】命令清空窗口。然后单击 Home 按钮显示桌面，可以查看已经执行的生命周期方法名称（如图 5-12 所示）。

图 5-10 单击多任务键的效果

图 5-11 "Logcat"窗口输出的结果（4）

图 5-12 "Logcat"窗口输出的结果（5）

5.1.5 Activity 的启动模式

Activity 启动模式有 4 种：standard、singleTop、singleTask 和 singleInstance。启动模式需要在 AndroidManifest.xml 文件的 android:launchMode 属性中进行设置，默认值为 standard。

> **提示：任务栈**
>
> 任务栈是一种放置 Activity 实例的容器，使用先进后出的栈进行存储。因此 Activity 不支持重新排序，只能根据压栈和出栈操作更改 Activity 的顺序。
>
> 在启动 App 时，系统会创建一个新的任务栈存储默认启动的 Activity，然后启动的其他 Activity 会被压入将其启动的 Activity 任务栈中并在前台显示出来。单击返回按钮（底部三角形的虚拟按钮或全面屏的返回手势），前台显示的 Activity 就会出栈。单击 Home 按钮（底部圆形的虚拟按钮）回到桌面，再启动另一个 App，此时前一个 App 就被移到后台，其任务栈成为后台任务栈。而刚启动的 App 创建的任务栈被调到前台，成为前台任务栈，显示的 Activity 就是前台任务栈中的栈顶元素。

1. standard：标准模式

启动 standard 标准模式的 Activity 时，创建一个新的实例，放置在启动该 Activity 的任务栈顶部。

在 C0504 工程中，包含 MainActivity 和 StandardActivity（如图 5-13 所示）。在 AndroidManifest.xml 文件中，StandardActivity 设置的启动模式是 standard。运行 C0504 工程，启动默认项 MainActivity 后，单击相应的按钮依次启动 StandardActivity、StandardActivity 和 StandardActivity（如图 5-14 所示）。

```
<activity android:name=".StandardActivity" android:launchMode="standard" />
```

图 5-13　MainActivity 和 StandardActivity 的运行效果

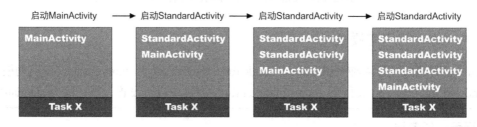

图 5-14　运行过程中任务栈的变化

在"Logcat"窗口中，可以看到每次启动的 StandardActivity 的任务栈 id 相同，但是 hashcode 都不相同（如图 5-15 所示）。此时单击 3 次返回按钮，才会返回 MainActivity，说明启动了 3 个 StandardActivity。

图 5-15　"Logcat"窗口输出的结果

> 　提示：standard 标准模式的使用场景
> 在没有特殊需求的情况下，大多使用标准模式启动 Activity。

2．singleTop：栈顶复用模式

启动 singleTop 栈顶复用模式的 Activity 处于当前栈的顶部，不会创建新的实例，而是直接启动该 Activity。onCreate()和 onStart()方法不会被调用，而是调用 onNewIntent()方法。当启动的 Activity 不在当前栈的顶部时，会创建一个新的实例。

在 C0505 工程中，包含 MainActivity 和 SingleTopActivity（如图 5-16 所示）。在 AndroidManifest.xml 文件中，SingleTopActivity 设置的启动模式是 singleTop。运行 C0505 工程，启动默认项 MainActivity 后，单击相应的按钮依次启动 SingleTopActivity、SingleTopActivity、MainActivity、SingleTopActivity 和 SingleTopActivity（如图 5-17 所示）。

```
<activity android:name=".SingleTopActivity" android:launchMode="singleTop" />
```

图 5-16　MainActivity 和 SingleTopActivity 的运行效果

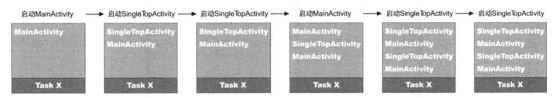

图 5-17　运行过程中任务栈的变化

在"Logcat"窗口中，可以观察到第二次启动 SingleTopActivity 后，没有调用 onCreate() 方法，而调用 onNewIntent()方法，输出的任务栈 id 和 hashcode 没有变化。第三次启动 SingleTopActivity 后，hashcode 发生了变化，说明创建了一个新实例。第四次启动 SingleTopActivity 后，没有调用 onCreate()方法，而调用 onNewIntent()方法，输出的任务栈 id 和 hashcode 与第三次启动的 SingleTopActivity 相同（如图 5-18 所示）。

图 5-18　"Logcat"窗口输出的结果

提示：singleTop 栈顶复用模式的使用场景

如果 Activity 在栈顶运行时，需要启动同类型的 Activity，使用该模式能够减少 Activity 实例的创建数量并节省内存。例如，在通知栏收到了三条新闻的推送信息，单击推送信息会启动显示新闻详情的 Activity。当单击第一条推送信息后，显示新闻详情的 Activity 已经处于栈顶。当单击第二条和第三条推送信息时，只需要通过 Intent 传入相应的数据即可，可以避免重复新建实例。

3. singleTask：栈内复用模式

任务栈中存在 singleTask 栈内复用模式的 Activity，当再次启动该 Activity 时，栈内该 Activity 上的所有 Activity 全部出栈，并且会回调该实例的 onNewIntent()方法。

在 C0506 工程中，包含 MainActivity 和 SingleTaskActivity（如图 5-19 所示）。在 AndroidManifest.xml 文件中，SingleTaskActivity 设置的启动模式是 singleTask。运行 C0506 工程，启动默认项 MainActivity 后，单击相应的按钮依次启动 SingleTaskActivity、MainActivity、MainActivity 和 SingleTaskActivity（如图 5-20 所示）。

```
<activity android:name=".SingleTaskActivity" android:launchMode="singleTask" />
```

图 5-19 MainActivity 和 SingleTaskActivity 的运行效果

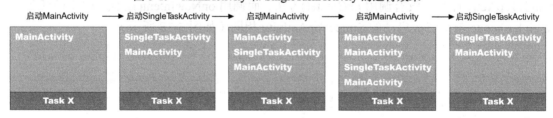

图 5-20 运行过程中任务栈的变化

在"Logcat"窗口中，可以观察到再次启动 SingleTaskActivity 时，调用 onNewIntent() 方法输出的任务栈 id 和 hashcode，与首次启动 SingleTaskActivity 时 onCreate() 方法输出的任务栈 id 和 hashcode 相同（如图 5-21 所示）。此时单击返回按钮会返回 MainActivity，再单击返回按钮会返回桌面。

图 5-21 "Logcat"窗口输出的结果

> 提示：singleTask 栈内复用模式的使用场景
> 栈内复用模式通常用于 App 首页的 Activity，且长时间保留在工作栈中，保证首页 Activity 的唯一性。

4. singleInstance：单实例模式

启动 singleInstance 单实例模式的 Activity 除了具有 singleTask 模式特性，还具有全局唯一性。系统中只能存在一个实例且单独占用一个任务栈，被该 Activity 开启的其他 Activity 会分配到其他任务栈内。首次启动该 Activity 时，会新建任务栈存储该 Activity 实例。再次启动该 Activity 时，除非该 Activity 实例已经销毁，否则不会在新栈内创建新的实例；重用该 Activity 实例，不会调用 onCreate() 方法，而调用 onNewIntent() 方法。

在 C0507 工程中，包含 MainActivity 和 SingleInstanceActivity（如图 5-22 所示）。在 AndroidManifest.xml 文件中，SingleInstanceActivity 设置的启动模式是 singleInstance。运行 C0507 工程，启动默认项 MainActivity 后，单击相应的按钮依次启动 SingleInstanceActivity、MainActivity 和 SingleInstanceActivity（如图 5-23 所示）。

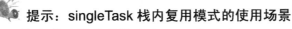

图 5-22 MainActivity 和 SingleInstanceActivity 的运行效果

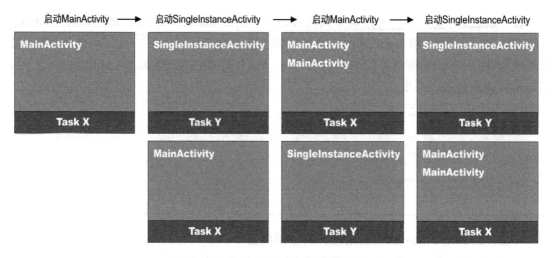

图 5-23 运行过程中任务栈的变化

在"Logcat"窗口中,可以观察到启动 SingleInstanceActivity 时,为其创建了新的任务栈。再次启动 MainActivity,被放置在原有的任务栈内,而不是 SingleInstanceActivity 所在的任务栈内。再次启动 SingleInstanceActivity,没有调用 onCreate()方法,调用 onNewIntent()方法输出的任务栈 id 和 hashcode,与第一次启动 SingleInstanceActivity 时调用 onCreate()方法输出的任务栈 id 和 hashcode 相同(如图 5-24 所示)。

图 5-24 "Logcat"窗口输出的结果

 提示:singleInstance 单实例模式的使用场景
单实例模式通常用于工具类的 App,被其他 App 调用时能够保证全局唯一性,如拨号、短信、相机、地图等。

5.1.6 实例工程:Activity 的数据传递

本实例演示了简单数据类型的数据通过 Intent 对象在各 Activity 之间进行传递(如图 5-25 所示),Intent 类的 putExtra()方法提供了丰富的重载方法,可以传递不同类型的数据。关闭 Activity 时,还可以回传数据。

1. 新建工程

新建一个"Empty Activity"工程,工程名称为"C0508"。

图 5-25　运行效果

2. 主界面的 Activity

```
/java/com/vt/c0508/MainActivity.java
09  public class MainActivity extends AppCompatActivity {
10      private static final int REQUEST_CODE = 101;
11      private EditText mResultEditText;
12
13      @Override
14      protected void onCreate(Bundle savedInstanceState) {
15          super.onCreate(savedInstanceState);
16          setContentView(R.layout.activity_main);
17          mResultEditText = findViewById(R.id.edit_text_result);
18          //发送数据（没有回传数据）
19          findViewById(R.id.button_send).setOnClickListener(new View.OnClickListener() {
20              @Override
21              public void onClick(View v) {
22                  Intent intent = new Intent(MainActivity.this, ReceiveActivity.class);
23                  intent.putExtra("id", 10407);
24                  intent.putExtra("msg", "MainActivity发送的信息1");
25                  startActivity(intent);
26                  mResultEditText.setText("");
27              }
28          });
29          //发送数据（接收回传数据）
30          findViewById(R.id.button_send_for_result).setOnClickListener(new View.OnClickListener(){
31              @Override
32              public void onClick(View view) {
33                  Intent intent = new Intent(MainActivity.this, ReceiveActivity.class);
34                  intent.putExtra("id", 20408);
35                  intent.putExtra("msg", "MainActivity发送的信息2");
```

```
36              startActivityForResult(intent, REQUEST_CODE);
37              mResultEditText.setText("");
38          }
39      });
40  }
41  //处理回传数据
42  protected void onActivityResult(int requestCode, int resultCode, Intent data) {
43      super.onActivityResult(requestCode, resultCode, data);
44      mResultEditText.setText("正在处理回传数据");
45      //判断请求码
46      if (requestCode == REQUEST_CODE) {
47          //判断返回码
48          if (resultCode == ReceiveActivity.RESULT_CODE) {
49              String result = data.getStringExtra("data");
50              mResultEditText.setText(result);
51          }else{
52              mResultEditText.setText("没有回传数据");
53          }
54      }
55  }
56  }
```

第 10 行使用静态整型常量 REQUEST_CODE 定义回传数据的请求码。第 19~28 行"发送数据（没有回传数据）"按钮的单击事件，使用 putExtra()方法将传递的数据存储在 intent 实例中，使用 startActivity(intent)启动 ReceiveActivity。第 30~40 行"发送数据（接收回传数据）"按钮的单击事件，同样使用 putExtra()方法将传递的数据存储在 intent 实例中，使用 startActivityForResult(intent, REQUEST_CODE) 启动 ReceiveActivity。第 42~55 行 onActivityResult(int requestCode, int resultCode, Intent data)方法根据 requestCode 和 resultCode 判断如何处理回传数据。第 48 行调用 ReceiveActivity 的静态整型常量 RESULT_CODE 判断是否 ReceiveActivity 使用该常量作为返回码回传数据。

3. 接收界面的 Activity

```
/java/com/vt/c0508/ReceiveActivity.java
09  public class ReceiveActivity extends AppCompatActivity {
10      public static final int RESULT_CODE = 201;
11      private EditText mIdEditText;
12      private EditText mMsgEditText;
13
14      @Override
15      protected void onCreate(Bundle savedInstanceState) {
16          super.onCreate(savedInstanceState);
17          setContentView(R.layout.activity_receive);
18          //获取传递数据
19          Intent intent = getIntent();
20          int id = intent.getIntExtra("id", 0);
21          String msg = intent.getStringExtra("msg");
22          //显示传递数据
23          mIdEditText = findViewById(R.id.edit_text_id);
```

```
24          mIdEditText.setText(String.valueOf(id));
25          mMsgEditText = findViewById(R.id.edit_text_name);
26          mMsgEditText.setText(msg);
27          //关闭（没有回传数据）
28          findViewById(R.id.button_finish).setOnClickListener(new View.OnClickListener(){
29              @Override
30              public void onClick(View view) {
31                  //关闭当前Activity
32                  finish();
33              }
34          });
35          //关闭（发送回传数据）
36          findViewById(R.id.button_finish_result).setOnClickListener(new View.OnClickListener(){
37              @Override
38              public void onClick(View view) {
39                  //设置返回的数据
40                  Intent intent = new Intent();
41                  intent.putExtra("data", "已经查阅信息！");
42                  setResult(RESULT_CODE, intent);
43                  //关闭当前Activity
44                  finish();
45              }
46          });
47      }
48  }
```

第10行使用静态整型常量RESULT_CODE定义回传数据的返回码。第28~34行"关闭（没有回传数据）"按钮的单击事件，直接使用finish()方法关闭当前Activity，不会向启动当前Activity的Activity回传数据。第36~46行"关闭（发送回传数据）"按钮的单击事件，通过intent实例存储回传数据，然后使用setResult(RESULT_CODE, intent)设置返回码和回传数据，最后使用finish()方法关闭当前Activity。

4．测试运行

运行C0508工程。在MainActivity中，单击"发送数据（没有回传数据）"按钮启动ReceiveActivity，ReceiveActivity显示接收的数据，再单击"关闭（没有回传数据）"或"关闭（发送回传数据）"按钮关闭ReceiveActivity，此时返回MainActivity并没有获取回传数据（如图5-26所示）。

在MainActivity中，单击"发送数据（接收回传数据）"按钮启动ReceiveActivity，ReceiveActivity显示接收的数据。当单击"关闭（没有回传数据）"按钮时，关闭ReceiveActivity返回MainActivity，此时虽然调用了onActivityResult(int requestCode, int resultCode, Intent data)方法，但是ReceiveActivity没有返回发送resultCode和data，无法获取有效的回传数据。当单击"关闭（发送回传数据）"按钮时，关闭ReceiveActivity返回MainActivity，调用了onActivityResult(int requestCode, int resultCode, Intent data)方法并获取回传数据（如图5-27所示）。

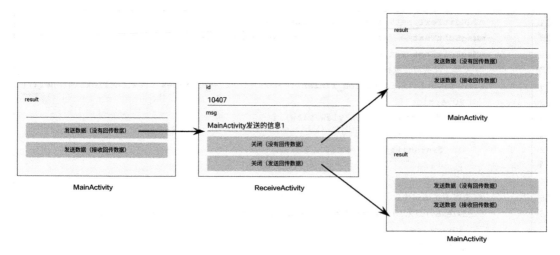

图 5-26 没有回传数据的效果

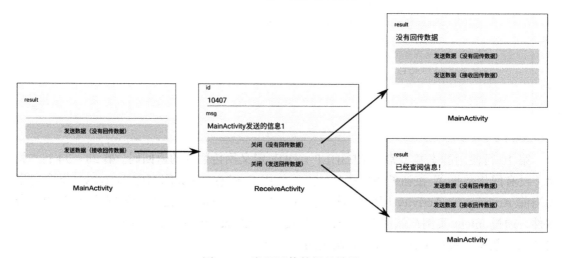

图 5-27 发送回传数据的效果

5.2 碎　　片

5.2.1 Fragment 概述

Fragment（androidx.fragment.app.Fragment）是嵌入 Activity 的程序单元。Fragment（碎片）类的常用方法如表 5-3 所示，其必须依赖于 Activity，不能独立存在。但因其有独立的生命周期，能接收输入事件，Activity 可以动态添加或删除 Fragment。一个 Activity 中可以包含多个 Fragment，一个 Fragment 可以被多个 Activity 重用。与 FragmentManager 类（如表 5-4 所示）和 FragmentTransaction 类（如表 5-5 所示）配合使用可以实现丰富的应用效果。

表 5-3　Fragment 类的常用方法

类型和修饰符	方　　法
Final FragmentActivity	getActivity() 获取与 Fragment 关联的 FragmentActivity。如果不关联，则返回 null
Context	getContext() 获取与 Fragment 关联的 Context
void	onActivityCreated(Bundle savedInstanceState) 当 Fragment 关联的 Activity 被创建且 Fragment 已经被实例化时，调用该方法
void	onActivityResult(int requestCode, int resultCode, Intent data) 接收前一次从 startActivityForResult(Intent, int) 调用的结果
void	onAttach(Context context) 当 Fragment 首次附加到它的 context 时，调用该方法
void	onAttachFragment(Fragment childFragment) 当一个 Fragment 附加到当前 Fragment 作为子对象时，调用该方法
void	onCreate(Bundle savedInstanceState) 当 Fragment 初始创建时，调用该方法
View	onCreateView(LayoutInflater inflater, ViewGroup container, Bundle savedInstanceState) 当 Fragment 实例化其用户界面视图时，调用该方法
void	onDestroy() 当 Fragment 销毁时，调用该方法
void	onDetach() 当 Fragment 不再附属于其 Activity 时，调用该方法
void	onHiddenChanged(boolean hidden) 当 Fragment 的隐藏状态改变，调用该方法
void	onInflate(Context context, AttributeSet attrs, Bundle savedInstanceState) 当 Fragment 被创建作为布局视图的一部分时，调用该方法
void	onPause() 当 Fragment 不再是 resumed 状态时，调用该方法
void	onPictureInPictureModeChanged(boolean isInPictureInPictureMode) 当 Fragment 关联的 Activity 改变画中画模式时，调用该方法
void	onResume() 当 Fragment 运行呈现给用户时，调用该方法
void	onSaveInstanceState(Bundle outState) 当要求 Fragment 保存当前动态状态时，调用该方法
void	onStart() 当 Fragment 呈现给用户时，调用该方法
void	onStop() 当 Fragment 不再是 started 状态时，调用该方法
void	onViewCreated(View view, Bundle savedInstanceState) 在 onCreateView(LayoutInflater, ViewGroup, Bundle)返回后，且在恢复任何已保存的状态到视图前，调用该方法
void	onViewStateRestored(Bundle savedInstanceState) 当所有已保存状态恢复到 Fragment 的视图层级体系后，调用该方法
void	startActivity(Intent intent) 从 Fragment 所包含的 Activity 调用 Activity.startActivity(Intent)方法
void	startActivity(Intent intent, Bundle options) 从 Fragment 所包含的 Activity 调用 Activity.startActivity(Intent, Bundle)方法
void	startActivityForResult(Intent intent, int requestCode) 从 Fragment 所包含的 Activity 调用 Activity.startActivityForResult(Intent, int)方法

表 5-4　FragmentManager 类的常用方法

类型和修饰符	方　法
FragmentTransaction	beginTransaction() 对与 FragmentManager 关联的所有 Fragment 开启一系列操作
int	getBackStackEntryCount() 返回当前在后堆栈中加入的数量
List<Fragment>	getFragments() 获取添加到 FragmentManager 中的 Fragment 列表

表 5-5　FragmentTransaction 类的常用方法

类型和修饰符	方　法
final FragmentTransaction	add(int containerViewId, Fragment fragment, String tag) 添加 Fragment
FragmentTransaction	attach(Fragment fragment) 在先前使用 detach(Fragment)从 UI 分离 Fragment 后，重新附加该 Fragment
abstract int	commit() 提交事务
abstract int	commitNow() 同步提交事务
FragmentTransaction	detach(Fragment fragment) 从 UI 中分离 Fragment
FragmentTransaction	hide(Fragment fragment) 隐藏指定的 Fragment
boolean	isEmpty() 判断是否有需要提交的事务
FragmentTransaction	remove(Fragment fragment) 移除指定的 Fragment
FragmentTransaction	replace(int containerViewId, Fragment fragment, String tag) 替换已有的 Fragment 到指定容器
FragmentTransaction	setCustomAnimations(int enter, int exit) 设置特定的动画资源作为进入和退出动画
FragmentTransaction	show(Fragment fragment) 显示之前隐藏的 Fragment

5.2.2　Fragment 的生命周期

Fragment 是依赖 Activity 使用的，Fragment 的生命周期和 Activity 的生命周期是有对应关系的（如图 5-28 所示）。

5.2.3　实例工程：导航分页的主界面

本实例演示了常用的带导航功能的主界面，底部包含 5 个导航按钮，第 5 个按钮上默认显示数字圆点提示（如图 5-29 所示）。单击导航按钮后，显示数字圆点提示会消失，顶部的标题会相应改变，并且中间的区域会显示相应的 Fragment。每个 Fragment 都包含一个"显示 TAB 圆点"按钮，单击后相应的导航按钮会显示数字圆点提示。

第 5 章　基本程序单元

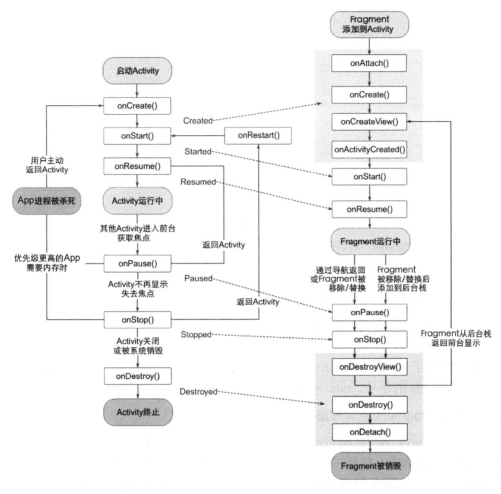

图 5-28　Fragment 和 Activity 的生命周期关系

1．打开基础工程

打开"基础工程"文件夹中的"C0509"工程，该工程已经包含 MainActivity 及布局的文件，为了使用更加灵活，导航没有使用 FragmentTabHost。

2．OnShowTabNumListener 接口

右键单击 "/java/com/vt/c0509" 文件夹，选择【New】→【Package】命令。在打开的 "New Package" 对话框中，输入包名，单击 "OK" 按钮完成创建（如图 5-30 所示）。

右键单击 "/java/com/vt/c0509/fragment" 文件夹，选择【New】→【Java Class】命令。在打开的 "New Java Class" 对话框中，设置 "Name" 为 "OnShowTabListener"，"Kind" 为 "Interface"，单击 "OK" 按钮（如图 5-31 所示）。

图 5-29　最终工程运行效果

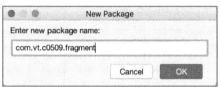
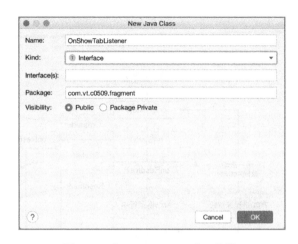

图 5-30 "New Package"对话框　　　图 5-31 "New Java Class"对话框

```
/java/com/vt/c0509/fragment/OnShowTabListener.java
03    public interface OnShowTabNumListener {
04        /*定义接口方法*/
05        void onShowTabNum(int index, int num);
06    }
```

第 05 行定义了 OnShowTabListener 接口的 onShowTabNum(int index, int num)。其中，index 表示 Fragment 的序号，num 表示图标上显示的提示数。

3．分页的 Fragment 及其布局

右键单击"/java/com/vt/c0509/fragment"文件夹，选择【New】→【Fragment】→【Fragment(Blank)】命令。打开"New Android Component"对话框，设置"Fragment Name"为"VicinityFragment"，取消勾选"Create layout XML?"和"Include fragment factory methods?"复选框，单击"Finish"按钮（如图 5-32 所示）。

图 5-32 "New Android Component"对话框

按照相同的方法新建 FollowFragment、MeetFragment、RemindFragment 和 MineFragment，将这些 Fragment 的布局设置成相同的效果。下面的布局代码仅展示 fragment_vicinity.xml 文件，其他 Fragment 布局文件的代码与其基本相同。

```xml
/res/layout/fragment_ vicinity.xml
01  <?xml version="1.0" encoding="utf-8"?>
02  <LinearLayout xmlns:android="http://schemas.android.com/apk/res/android"
03      xmlns:tools="http://schemas.android.com/tools"
04      android:layout_width="match_parent"
05      android:layout_height="match_parent"
06      android:gravity="center"
07      android:orientation="vertical"
08      tools:context=".fragment.VicinityFragment">
09      <TextView
10          android:layout_width="wrap_content"
11          android:layout_height="wrap_content"
12          android:text="VicinityFragment"
13          android:textSize="28sp" />
14      <Button
15          android:id="@+id/button"
16          android:layout_width="match_parent"
17          android:layout_height="wrap_content"
18          android:text="显示 tab 圆点" />
19  </LinearLayout>
```

在新建的 4 个 Fragment 中，MeetFragment 使用内部定义接口与 MainActivity 传递数据，其他 Fragment 使用外部接口与 MainActivity 传递数据。

```java
/java/com/vt/c0509/fragment/VicinityFragment.java
11  public class VicinityFragment extends Fragment {
12      @Override
13      public View onCreateView(LayoutInflater inflater, ViewGroup container, Bundle savedInstanceState) {
14          View view = inflater.inflate(R.layout.fragment_vicinity, container, false);
15          Button button = view.findViewById(R.id.button);
16          button.setOnClickListener(new View.OnClickListener() {
17              @Override
18              public void onClick(View v) {
19                  //((MainActivity) getActivity()).showTabNum(4, 5);
20                  if (getContext() instanceof OnShowTabNumListener) {
21                      OnShowTabNumListener onShowTabNumListener;
22                      onShowTabNumListener = (OnShowTabNumListener) getContext();
23                      onShowTabNumListener.onShowTabNum(0, 12);
24                  } else {
25                      throw new RuntimeException(getContext().toString()
26                              + " must implement OnShowTabNumListener");
27                  }
28              }
29          });
30          return view;
31      }
32  }
```

第 19 行注释的代码可以直接调用 MainActivity 的 showTabNum()方法。第 20 行判断 Fragment 关联的 Activity 是不是 OnShowTabNumListener 接口的实例,没有关联的 Activity 无法调用关联的 Activity 中的接口方法。第 21～23 行新建 onShowTabNumListener 实例,调用 OnShowTabNumListener 接口的 onShowTabNum()方法显示圆点提示。第 25 行当无法获取到关联的 Activity 时,抛出自定义异常。

```
/java/com/vt/c0509/fragment/MeetFragment.java
11  public class MeetFragment extends Fragment {
12      @Override
13      public View onCreateView(LayoutInflater inflater, ViewGroup container, Bundle
14  savedInstanceState) {
15          View view = inflater.inflate(R.layout.fragment_meet, container, false);
16          Button button = view.findViewById(R.id.button);
17          button.setOnClickListener(new View.OnClickListener() {
18              @Override
19              public void onClick(View v) {
20                  //((MainActivity) getActivity()).showTabNum(2, 3);
21                  if (getContext() instanceof OnShowTabNumListener) {
22                      OnShowTabNumListener onShowTabNumListener;
23                      onShowTabNumListener = (OnShowTabNumListener) getContext();
24                      onShowTabNumListener.onShowTabNum(2, 5);
25                  } else {
26                      throw new RuntimeException(getContext().toString()
27                              + " must implement OnShowTabNumListener");
28                  }
29              }
30          });
31          return view;
32      }
33      //内部接口
34      public interface OnShowTabNumListener {
35          //定义接口方法
36          void onShowTabNum(int index, int num);
37      }
38  }
```

第 33～36 行定义了一个内部接口 OnShowTabNumListener,该接口名称和定义的接口方法与 Fragment 包内的 OnShowTabNumListener 接口名称和定义的接口方法相同。

4. 主界面的 Activity

在 MainActivity 中实现对主界面导航的控制,并且实现了 OnShowTabNumListener 和 MeetFragment.OnShowTabNumListener 的接口方法。

```
/java/com/vt/c0509/MainActivity.java
12  public class MainActivity extends AppCompatActivity implements OnShowTabNumListener,
    MeetFragment.OnShowTabNumListener {
13      private TextView mTitleTextView;
14      private FragmentManager mFragmentManager;
15      private String[] mTitles = new String[]{"附近", "关注", "偶遇", "提醒", "自己"};
```

```
16        private TextView[] mTab = new TextView[5];
17        private int[] mTabId = new int[]{R.id.tab_vicinity, R.id.tab_follow, R.id.tab_meet,
   R.id.tab_remind, R.id.tab_mine};
18        private int[] mTabNumId = new int[]{R.id.tab_vicinity_num, R.id.tab_follow_num,
   R.id.tab_meet_num, R.id.tab_remind_num, R.id.tab_mine_num};
19        private TextView[] mTabNum = new TextView[5];
20        private Fragment[] mFragment = new Fragment[]{new VicinityFragment(), new
   FollowFragment(), new MeetFragment(), new RemindFragment(), new MineFragment()};
21
22        @Override
23        protected void onCreate(Bundle savedInstanceState) {
24            super.onCreate(savedInstanceState);
25            setContentView(R.layout.activity_main);
26            //初始化控件
27            mTitleTextView = findViewById(R.id.text_view_title);
28            for (int i = 0; i < mTab.length; i++) {
29                mTab[i] = findViewById(mTabId[i]);
30                mTab[i].setTag(i);
31                mTab[i].setOnClickListener(new View.OnClickListener() {
32                    @Override
33                    public void onClick(View v) {
34                        showTabFragment((int) v.getTag());
35                    }
36                });
37                mTabNum[i] = findViewById(mTabNumId[i]);
38            }
39            //必须使用 FragmentActivity.getSupportFragmentManager()获取 FragmentManager 实例
40            mFragmentManager = getSupportFragmentManager();
41            showTabFragment(2);//设置默认显示的 Fragment
42            showTabNum(4, 5);//设置 Tab 显示圆点数字
43        }
44        //隐藏 Tab 的圆点数字
45        private void hideTabNum(int index) {
46            mTabNum[index].setVisibility(View.GONE);
47        }
48        //显示 Tab 的圆点数字
49        public void showTabNum(int index, int num) {
50            mTabNum[index].setText(String.valueOf(num));
51            mTabNum[index].setVisibility(View.VISIBLE);
52        }
53        //显示 Fragment
54        private void showTabFragment(int index) {
55            //创建事务
56            FragmentTransaction fragmentTransaction = mFragmentManager.beginTransaction();
57            //判断是否已经添加 Fragment
58            if (mFragmentManager.getFragments().isEmpty()) {
59                for (int i = 0; i < mTab.length; i++) {
60                    fragmentTransaction.add(R.id.fragment_content, mFragment[i]);//添加 Fragment
61                }
62            }
```

```
63              //还原状态
64              for (int i = 0; i < mTab.length; i++) {
65                  fragmentTransaction.hide(mFragment[i]);//隐藏 Fragment
66                  mTab[i].setSelected(false);//取消 Tab 选中状态
67              }
68              mTitleTextView.setText(mTitles[index]);//设置标题
69              mTab[index].setSelected(true);//设置 Tab 选中状态
70              hideTabNum(index);//隐藏 Tab 圆点
71              fragmentTransaction.show(mFragment[index]);//显示 Fragment
72              fragmentTransaction.commit();//提交事务
73          }
74          //实现接口方法
75          @Override
76          public void onShowTabNum(int index, int num) {
77              showTabNum(index, num);
78          }
79      }
```

第15~20行使用数组创建实例初始化导航功能所需的数据,这里使用数组的优势在于可以很容易地增减导航菜单并减少代码量。第28~38行使用 for 循环遍历数组的所有元素,初始化每个导航按钮。第41行设置默认选中的导航按钮并显示相应的 Fragment。第42行设置第5个导航按钮显示圆点提示,圆点内的数字是5,这行代码通常应该在从服务器端获取未读的信息数量后使用,这里只是一个模拟效果。第58~62行判断是否已经向 mFragmentManager 添加过 Fragment,因为重复通过 fragmentTransaction.add()方法向 mFragmentManager 添加 Fragment 会报错。第72行 fragmentTransaction.commit()提交事务才会执行 add()、hide()和 show()方法的事务。第75~78行实现了接口的 onShowTabNum 方法。

5.3 习 题

1. 实现登录后显示用户名的效果,输入用户名和密码登录,启动一个新的 Activity 显示用户名。

2. 以 C0402 工程为基础,单击图像启动一个单独显示图像并包含"删除"按钮的 Activity,单击"删除"按钮关闭 Activity 并移除该列表项。

3. 使用 Fragment 实现"小红书"的导航效果,底部导航包含首页、商城、+、消息和我。

第 6 章 后台服务与广播

Activity 和 Fragment 不在前台显示时是暂停运行的,此时如果有些功能需继续运行,就要通过后台服务来实现,后台服务甚至在关闭 App 时还可以在后台继续运行。而广播为 App 之间提供了一种单向数据传递的方式,能够实现一对多的实时数据传递。其中,系统广播是系统根据设备的各种状态发送的广播。

6.1 服 务

6.1.1 Service 概述

Service(android.app.Service)是在后台可以长时间运行且没有可视化视图的组件,Service(服务)类的常用方法如表6-1所示。Service 不是线程,也不是在主线程外的方法。使用 Context.startService(Intent)方法可以启动服务,使用 Context.stopService(Intent)方法可以停止服务。

表 6-1 Service 类的常用方法

类型和修饰符	方　　法
final Application	getApplication() 获取 Service 所属的 Application
abstract IBinder	onBind(Intent intent) 绑定后调用该方法,返回 IBinder 实例,用于与客户端进行数据通信。没有绑定到服务时,可以返回 null
void	onCreate() 创建时调用该方法
void	onDestroy() 被销毁前调用该方法
void	onRebind(Intent intent) 新客户端绑定时调用该方法,在此之前已通知其 onUnbind(Intent)中的所有绑定解除
int	onStartCommand(Intent intent, int flags, int startId) 当调用 Context.startService(Intent)方法启动服务时,调用该方法
void	onTaskRemoved(Intent rootIntent) 当 Service 运行,用户移除 Service 所属的 App 任务时,调用该方法
boolean	onUnbind(Intent intent) 当所有客户端都解除绑定时,调用该方法
final void	startForeground(int id, Notification notification) 如果 Service 已通过 Context.startService(Intent)启动,则持续显示 Notification
final void	stopForeground(boolean removeNotification) 如果 removeNotification 为 true,则删除 Notification

续表

类型和修饰符	方 法
final void	stopSelf() 如果 Service 已经开始运行，则停止 Service
final boolean	stopSelfResult(int startId) 停止使用 startId 启动的 Service

Service 需要在 AndroidManifest.xml 文件中添加<service>标签后才能被调用，<service>标签属性（如表 6-2 所示）大多无法通过 Service 类的方法进行设置和获取。

表 6-2 <service>标签属性

属 性	说 明
android:description	设置描述 Service 的字符串
android:directBootAware	设置是否可以在用户解锁前直接启动
android:enabled	设置是否可以实例化 Service
android:exported	设置是否允许其他 App 调用该服务
android:name	设置实现 Service 的类名
android:process	设置运行 Service 的进程名称。如果名称以":"开头，则在需要时创建专用新进程运行；如果名称以小写字母开头，则在使用该名称的全局进程中运行

使用 Context.bindService(Intent i, ServiceConnection conn, int flags)方法可以绑定服务，flags 使用 Service 类中与绑定相关的常量（如表 6-3 所示）。绑定的 Service 会随着 App 的退出而终止，未绑定的 Service 不会随着 App 的退出而终止。使用 Context.unbindService(ServiceConnection)方法可以解除绑定。

表 6-3 Service 类中与绑定相关的常量

常 量	说 明
BIND_AUTO_CREATE	绑定后自动创建服务，无须使用 startService(Intent)方法后启动
BIND_ADJUST_WITH_ACTIVITY	如果从某个 Activity 绑定，则允许根据该 Activity 是否对用户可见来提高目标 Service 进程的重要性，不论是否使用另一个标志来减少使用客户端流程的总体重要性从而影响它的数量
BIND_ABOVE_CLIENT	指示绑定 Service 的客户端比 App 本身更重要。当内存不足时，系统在结束 Service 之前先终止应用程序
BIND_ALLOW_OOM_MANAGEMENT	允许对绑定 Service 进行内存溢出管理。当需要内存时，允许系统结束绑定 Service 的进程
BIND_DEBUG_UNBIND	设置此标志，将保留 unbindService(ServiceConnection)调用的堆栈；如果稍后发生了不正确的解除绑定调用，则会打印该堆栈。建议只用于调试
BIND_EXTERNAL_SERVICE	绑定的 Service 是一个独立的外部服务
BIND_IMPORTANT	此服务对客户端非常重要，在客户端处于前台进程级别时，应将其带到前台进程级别
BIND_INCLUDE_CAPABILITIES	如果绑定的 App 由于其前台状态（如 Activity 或前台 Service）而具有特定功能，只要 App 具有所需的权限，就允许绑定的 App 获取相同的功能
BIND_NOT_FOREGROUND	不允许此绑定提升 Service 进程到前台优先级，但仍会被提升到不高于客户端的优先级

续表

常量	说明
BIND_NOT_PERCEPTIBLE	如果绑定来自可见或用户可感知的 App，则降低 Service 的重要性到可感知级别以下
BIND_WAIVE_PRIORITY	不要影响 Service 主进程的调度或内存管理优先级

6.1.2 Service 的生命周期

Service 分为非绑定 Service 和绑定 Service，二者生命周期有所不同（如图 6-1 所示）。非绑定 Service 包含 3 个生命周期方法，绑定 Service 包含 4 个生命周期方法。

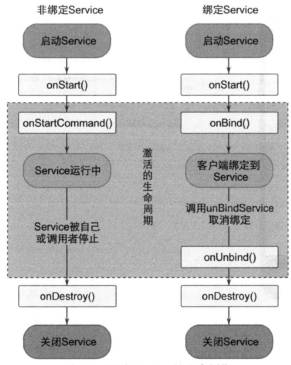

图 6-1 两种 Service 的生命周期

两种 Service 都会调用 onCreate()方法来创建 Service，但绑定 Service 不会调用 onStartCommand()方法，而是调用 onBind()方法返回客户端一个 IBinder 接口。绑定 Service 通过 Context.unbindService()方法解除绑定时，回调 onUnbind()方法。当所有绑定 Service 都调用了 unbindService()方法，绑定 Service 会被停止，然后回调 onDestroy() 方法。

6.1.3 实例工程：Service 的开启和停止

本实例演示了启动和停止 Service 的方法（如图 6-2 所示），单击"启动 SERVICE"按钮启动 MyService 类实例在后台运行，单击"停止 SERVICE"按钮停止 MyService 类实例的运行。

1. 打开基础工程

打开"基础工程"文件夹中的"C0601"工程，该工程已经包含 MainActivity 及布局的文件。

2. 新建 Service

选择"/java/com/vt/c0601"文件夹，单击右键并选择【New】→【Service】→【Service】命令。在打开的"New Android Component"对话框中，设置"Class Name"为"MyService"，单击"Finish"按钮（如图 6-3 所示）。

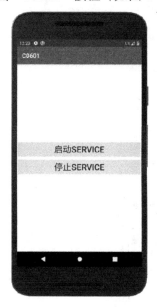
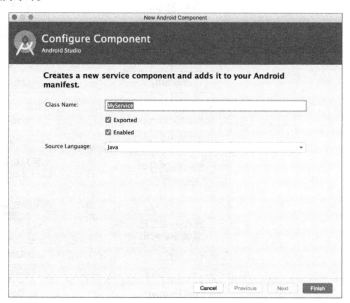

图 6-2　运行效果　　　　图 6-3　"New Android Component"对话框

此时，需要在 AndroidManifest.xml 文件的<application>标签中添加<service>标签声明服务，否则无法使用。

```
/manifests/AndroidManifest.xml
12    <service
13        android:name=".MyService"
14        android:enabled="true"
15        android:exported="true"/>
```

第 13 行 android:name 属性设置服务的类名。第 14 行 android:enabled 属性表示是否可以使用，默认值为 true；当父标签的 enabled 属性也为 true 时，服务才会被激活，否则不会激活。第 15 行 android:exported 属性表示其他 App 的组件是否可以唤醒 Service 或与 Service 进行交互，默认值为 true。

```
/java/com/vt/c0601/MyService.java
08    public class MyService extends Service {
09        private final String TAG = "MyService";
10        private boolean mServiceRunning = false;
```

```
11      private int mSecond = 0;
12      //Service绑定时调用该方法，必须要实现的Service类接口方法
13      @Override
14      public IBinder onBind(Intent intent) {
15          Log.i(TAG, "onBind方法被调用!");
16          return null;
17      }
18      //Service被创建时调用
19      @Override
20      public void onCreate() {
21          Log.i(TAG, "onCreate方法被调用!");
22          super.onCreate();
23      }
24      //Service被启动时调用
25      @Override
26      public int onStartCommand(Intent intent, int flags, int startId) {
27          Log.i(TAG, "onStartCommand方法被调用!");
28          new Thread() {
29              @Override
30              public void run() {
31                  mServiceRunning = true;
32                  while (mServiceRunning) {
33                      Log.i(TAG, "新线程已经运行了" + mSecond + "秒");
34                      try {
35                          sleep(1000);
36                          mSecond++;
37                      } catch (InterruptedException e) {
38                          e.printStackTrace();
39                      }
40                  }
41              }
42          }.start();
43          return super.onStartCommand(intent, flags, startId);
44      }
45      //Service被关闭前回调
46      @Override
47      public void onDestroy() {
48          Log.i(TAG, "onDestroy方法被调用!");
49          mServiceRunning = false;
50          super.onDestroy();
51      }
52  }
```

第28～42行创建新线程在后台持续运行，通过mServiceRunning判断是否结束循环结束线程。第46～51行onDestroy()方法在Service被关闭前调用，将mServiceRunning值修改为false，用于关闭新线程。

3．主界面的Activity

```
/java/com/vt/c0601/MainActivity.java
09  public class MainActivity extends AppCompatActivity {
```

```
10      private final String TAG = "MainActivity";
11      @Override
12      protected void onCreate(Bundle savedInstanceState) {
13          super.onCreate(savedInstanceState);
14          setContentView(R.layout.activity_main);
15          //创建启动Service的Intent及Intent属性
16          final Intent intent = new Intent(this, MyService.class);
17          findViewById(R.id.button_start).setOnClickListener(new View.OnClickListener() {
18              @Override
19              public void onClick(View v) {
20                  startService(intent);//启动Service
21                  Log.i(TAG, "startService方法被调用");
22              }
23          });
24          findViewById(R.id.button_stop).setOnClickListener(new View.OnClickListener() {
25              @Override
26              public void onClick(View v) {
27                  stopService(intent);//停止Service
28                  Log.i(TAG, "stopService方法被调用!");
29              }
30          });
31      }
32  }
```

第 20 行使用 startService(intent)启动 Service。第 27 行使用 stopService(intent)停止 Service。

4．测试运行

运行后，单击"启动 SERVICE"按钮后等待几秒，再单击"停止 SERVICE"按钮，在"Logcat"窗口中观察输出结果（如图 6-4 所示）。

图 6-4 "Logcat"窗口输出的结果

6.1.4 实例工程：Service 的绑定和数据传递

本实例演示了 Service 的绑定和数据传递，以及关闭启动 Service 的 Activity 时如何停止 Service（如图 6-5 所示）。单击"启动 SubActivity"按钮启动 SubActivity。在 SubActivity 中，依次单击"启动 Service"和"绑定 Service"按钮后，再单击"获取 Service 运行时间"按钮会通过提示信息显示 Service 已经运行的时间。未停止 Service 时，关闭 SubActivity 会停止 Service。

图 6-5　运行效果

1．打开基础工程

打开"基础工程"文件夹中的"C0602"工程，该工程已经包含 MainActivity、SubActivity 类及布局的文件。

2．新建 Service

在工程中，选择"/java/com/vt/c0602"文件夹，单击右键并选择【New】→【Service】→【Service】命令。在打开的"New Android Component"对话框中，设置"Class Name"为"MyService"，单击"Finish"按钮。

```
/java/com/vt/c0602/MyService.java
09   public class MyService extends Service {
10       private final String TAG = "MyService";
11       private MyBinder mBinder = new MyBinder();
12       private boolean mServiceRunning = false;
13       private int mSecond;
14       //创建MyBinder类，继承自Binder，用于与Activity传递数据
15       public class MyBinder extends Binder {
16           public int getSecond() {
17               return mSecond;
18           }
19       }
20       //Service绑定时回调该方法，继承Service后该方法必须实现的方法
21       @Override
22       public IBinder onBind(Intent intent) {
23           Log.i(TAG, "onBind方法被调用!");
24           return mBinder;
25       }
26       //Service被创建时回调
27       @Override
```

```java
28    public void onCreate() {
29        super.onCreate();
30        Log.i(TAG, "onCreate方法被调用!");
31    }
32    //Service被启动时调用
33    @Override
34    public int onStartCommand(Intent intent, int flag, int startId) {
35        new Thread() {
36            public void run() {
37                mServiceRunning = true;
38                while (mServiceRunning) {
39                    Log.i(TAG, "新线程已经运行了" + mSecond + "秒");
40                    try {
41                        sleep(1000);
42                        mSecond++;
43                    } catch (InterruptedException e) {
44                        e.printStackTrace();
45                    }
46                }
47            }
48        }.start();
49        return super.onStartCommand(intent, flag, startId);
50    }
51    //Service重新绑定时回调
52    @Override
53    public void onRebind(Intent intent) {
54        super.onRebind(intent);
55        Log.i(TAG, "onRebind方法被调用!");
56    }
57    //Service解除绑定时回调
58    @Override
59    public boolean onUnbind(Intent intent) {
60        Log.i(TAG, "onUnbind方法被调用!");
61        mServiceRunning = false;
62        return true;
63    }
64    //Service被关闭前回调
65    @Override
66    public void onDestroy() {
67        super.onDestroy();
68        Log.i(TAG, "onDestroyed方法被调用!");
69        mServiceRunning = false;
70    }
71    //Service所属的App被移除时回调该方法
72    @Override
73    public void onTaskRemoved(Intent rootIntent) {
74        Log.i(TAG, "onTaskRemoved方法被调用!");
75    }
76 }
```

第 15~19 行创建 MyBinder 类，内部的 getSecond()方法获取运行时间。第 24 行的

mBinder 实例作为绑定的返回值，是传递数据的对象。第 34~50 行 Service 被启动后开启新线程，实现不精确计时的功能，统计 Service 已经运行的时间。

3. 控制 Service 的 Activity

```
/java/com/vt/c0602/MainActivity.java
14    public class SubActivity extends AppCompatActivity {
15        private final String TAG = "MainActivity";
16        private boolean mIsBind = false;
17        //保持所启动的 Service 的 IBinder 对象,同时定义一个 ServiceConnection 对象
18        private MyService.MyBinder mIBinder;
19        private ServiceConnection mConn = new ServiceConnection() {
20            //Activity 与 Service 连接成功时回调该方法
21            @Override
22            public void onServiceConnected(ComponentName name, IBinder iBinder) {
23                Log.i(TAG, "onServiceConnected 方法被调用!");
24                mIBinder = (MyService.MyBinder) iBinder;
25            }
26            //Activity 与 Service 断开连接时回调该方法
27            @Override
28            public void onServiceDisconnected(ComponentName name) {
29                Log.i(TAG, "onServiceDisconnected 方法被调用!");
30                mIBinder = null;
31            }
32        };
33        @Override
34        protected void onCreate(Bundle savedInstanceState) {
35            super.onCreate(savedInstanceState);
36            setContentView(R.layout.activity_sub);
37            this.setTitle("SubActivity");
38            //创建启动 Service 的 Intent
39            final Intent intent = new Intent(this, MyService.class);
40            findViewById(R.id.button_start).setOnClickListener(new View.OnClickListener() {
41                @Override
42                public void onClick(View v) {
43                    startService(intent);//启动 Service
44                }
45            });
46            findViewById(R.id.button_stop).setOnClickListener(new View.OnClickListener() {
47                @Override
48                public void onClick(View v) {
49                    stopService(intent);//停止 Service
50                }
51            });
52            findViewById(R.id.button_bind).setOnClickListener(new View.OnClickListener() {
53                @Override
54                public void onClick(View v) {
55                    bindService(intent, mConn, Service.BIND_IMPORTANT);//绑定 Service
56                    mIsBind = true;
57                }
```

```
58              });
59              findViewById(R.id.button_unbind).setOnClickListener(new View.OnClickListener() {
60                  @Override
61                  public void onClick(View v) {
62                      unbindService(mConn);//解除绑定 Service
63                      mIsBind = false;
64                  }
65              });
66              findViewById(R.id.button_get).setOnClickListener(new View.OnClickListener() {
67                  @Override
68                  public void onClick(View v) {
69                      if(mIBinder != null) {
70                          Toast.makeText(getApplicationContext(), "Service 已经运行了" +
     mIBinder.getSecond() + "秒!", Toast.LENGTH_SHORT).show();
71                      }
72                  }
73              });
74          }
75          @Override
76          protected void onDestroy() {
77              super.onDestroy();
78              Log.i(TAG, "onDestroy方法被调用!");
79              if(mIsBind) {
80                  unbindService(mConn);//解除绑定 Service
81              }
82          }
83      }
```

第 19～32 行实例化 ServiceConnection 对象，重写 onServiceConnected()和 onServiceDisconnected()方法。第 24 行将 MyService 的 onBind()方法返回值进行强制类型转换后赋值给 mIBinder。第 76～82 行重写 onDestroy()方法，如果在 Activity 销毁前已经绑定了 Service，则解除绑定。因为在 Activity 销毁时，如果 Activity 绑定了 Service 会报错，所以要解除绑定。

4．测试运行

运行后，单击"启动 SubActivity"按钮打开 SubActivity 后，依次单击"启动 Service"按钮、"绑定 Service"按钮、"解除绑定 Service"按钮和"停止 Service"按钮，再单击返回按钮关闭 SubActivity，在"Logcat"窗口中观察输出结果（如图 6-6 所示）。

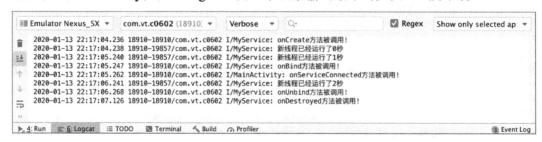

图 6-6 "Logcat"窗口输出的结果（1）

运行后，单击"启动 SubActivity"按钮打开 SubActivity，再依次单击"启动 Service"按钮、"绑定 Service"按钮和"停止 Service"按钮，在"Logcat"窗口中观察输出结果（如图 6-7 所示）。

图 6-7 "Logcat"窗口输出的结果（2）

运行后，单击"启动 SubActivity"按钮打开 SubActivity，再依次单击"启动 Service"按钮、"绑定 Service"按钮，单击返回按钮关闭 SubActivity；然后单击"启动 SubActivity"按钮打开 SubActivity 后，单击"启动 Service"按钮，在"Logcat"窗口中观察到输出结果（如图 6-8 所示）。可以看到：再次打开 SubActivity 后单击"启动 Service"按钮，输出的运行时间并没有重置，而是连续的，说明启动 Service 后没有使用 stopService(intent)方法停止该 Service，再次使用 startService(intent)启动该 Service 时该 Service 并未被销毁，因此不会执行 MyService 类的 onCreate()方法，但会再次执行 MyService 类的 onStartCommand(Intent, int, int)方法。

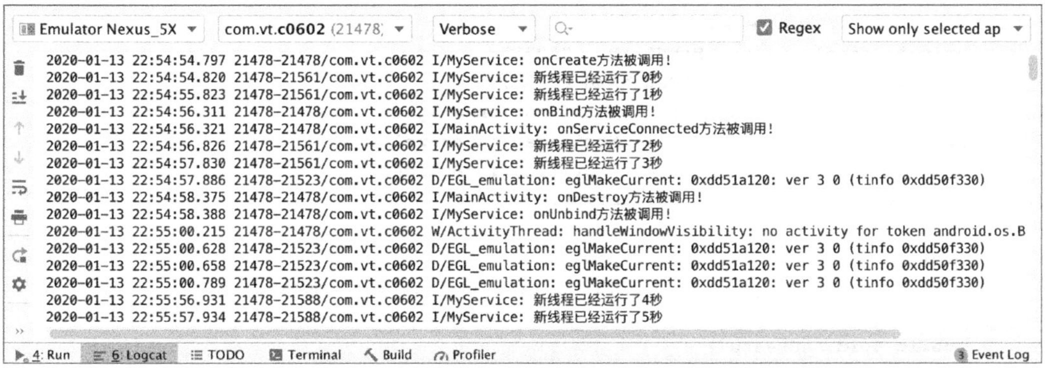

图 6-8 "Logcat"窗口输出的结果（3）

6.1.5 实例工程：Service 显示 Notification

本实例演示了使用 Service 显示前台通知（如图 6-9 所示）。单击"启动前台通知服务"按钮，在状态栏显示通知图标，将状态栏下拉后会看到通知。只有单击"停止前台通知服务"按钮才能关闭通知；否则，即使关闭该 App，通知依然不会消失。

1. 打开基础工程

打开"基础工程"文件夹中的"C0603"工程，该工程已经包含通知的布局文件及图标文件、MainActivity 及布局的文件。

图 6-9 运行效果

2．新建 Service

选择"/java/com/vt/c0603"文件夹，单击右键并选择【New】→【Service】→【Service】命令。在打开的"New Android Component"对话框中，设置"Class Name"为"MyService"，单击"Finish"按钮。

```
/java/com/vt/c0603/MyService.java
11  public class MyService extends Service {
12      private final String CHANNL_ID = "1231";
13      private final String CHANNL_NAME = "音乐";
14      private NotificationManager mNotificationManager;
15      private NotificationChannel mNotificationChannel;
16      private Notification.Builder mNotificationBuilder;
17      private Notification mNotification;
18      @Override
19      public IBinder onBind(Intent intent) {
20          throw new UnsupportedOperationException("Not yet implemented");
21      }
22      public void onCreate() {
23          super.onCreate();
24          //自定义远程视图
25          RemoteViews remoteView = new RemoteViews(getPackageName(), R.layout.notification_custom);
26          remoteView.setImageViewResource(R.id.image_view_music, R.mipmap.ic_notification_music);
27          remoteView.setImageViewResource(R.id.image_view_stop, R.mipmap.ic_notification_stop);
28          //设置通知
29          mNotificationChannel = new NotificationChannel(CHANNL_ID, CHANNL_NAME, NotificationManager.IMPORTANCE_DEFAULT);
30          mNotificationManager = (NotificationManager) getSystemService(NOTIFICATION_SERVICE);
31          mNotificationManager.createNotificationChannel(mNotificationChannel);
32          mNotificationBuilder = new Notification.Builder(this, CHANNL_ID);
33          mNotificationBuilder.setSmallIcon(R.mipmap.ic_notification_small)
34                  .setCustomBigContentView(remoteView);
35          mNotification = mNotificationBuilder.build();
36          //前台启动通知
```

```
37          startForeground(1, mNotification);
38      }
39  }
```

第 26、27 行设置远程视图的图像，布局文件中的图像需要重新设置才会显示出来。第 34 行设置通知的自定义大视图。第 37 行使用 startForeground() 方法启动前台通知。

```
/manifests/AndroidManifest.xml
03  <uses-permission android:name="android.permission.FOREGROUND_SERVICE" />
11      <service
12          android:name=".MyService"
13          android:enabled="true"
14          android:exported="true">
15      </service>
```

第 03 行添加前台服务的权限，否则 startForeground() 方法会报错。第 11~15 行是新建 MyService 后自动添加的标签。

3．主界面的 Activity

```
/java/com/vt/c0603/MainActivity.java
15  final Intent intent = new Intent(this, MyService.class);
16  findViewById(R.id.button_start_foreground).setOnClickListener(new
17  View.OnClickListener() {
18      @Override
19      public void onClick(View v) {
20          startService(intent);
21      }
22  });
23  findViewById(R.id.button_stop_foreground).setOnClickListener(new
24  View.OnClickListener() {
25      @Override
26      public void onClick(View v) {
27          stopService(intent);
        }
    });
```

第 16~21 行设置"启动前台通知服务"按钮启动 MyService。第 23~27 行设置"停止前台通知服务"按钮停止 MyService。

6.2 独立线程服务

6.2.1 IntentService 概述

IntentService（android.app.IntentService）是实现多线程处理异步请求的 Service 子类。IntentService 类的常用方法如表 6-4 所示。由于 Service 运行在主线程中，进行一些耗时的操作时会出现 ANR 问题，所以 IntentService 并不在主线程中运行，可以解决 ANR 问题。当然，Service 也可以在 onStartCommon(Intent, int, int) 方法中新建子线程执行耗时的操作，以解决 ANR 问题。

表 6-4　IntentService 类的常用方法

类型和修饰符	方　　法
	IntentService(String name) 构造方法
IBinder	onBind(Intent intent) 绑定后调用该方法，用于返回 IBinder 实例与客户端进行数据通信
void	onCreate() 服务首次被创建时调用该方法
void	onDestroy() 不再使用或移除时调用该方法
int	onStartCommand(Intent intent, int flags, int startId) 调用 Context.startService(Intent)方法启动服务时调用该方法
void	setIntentRedelivery(boolean enabled) 设置 Intent 是否重新传递，默认值为 false
protected abstract void	onHandleIntent(Intent intent) 处理 Intent 请求时调用该方法，该方法是被保护的方法，可以通过重写的方式使用

> **提示：ANR**
>
> ANR（Application Not Responding）是指 App 无响应。当 App 的 UI 线程响应超时才会引起 ANR，超时产生的原因一般有两种：当前的事件没有机会得到处理，如 UI 线程正在响应另外一个事件，当前事件由于某种原因被阻塞了；当前的事件正在处理，但是由于耗时太长没有及时完成。从本质上讲，产生 ANR 的原因有三种，大致对应四大组件中的三个（Activity/View，BroadcastReceiver 和 Service）。
> - KeyDispatchTimeOut：最常见的一种类型，原因是 View 的按键事件或触摸事件在特定时间（如 5s）内无法得到响应。
> - BroadcastTimeOut: 原因是 BroadcastReceiver 的 onReceive()函数运行在主线程中，在特定时间（如 10s）内无法完成处理。
> - ServiceTimeOut: 较少出现的一种类型，原因是 Service 的各个生命周期函数在特定时间（如 20s）内无法完成处理。

6.2.2　实例工程：IntentService 的静态方法启动

本实例演示了使用 IntentService 自定义类的静态方法进行启动，并根据启动时传入的参数判断进行加法还是乘法运算（如图 6-10 所示）。单击"启动 add 动作的 IntentService"按钮，开启新线程，运行一次加法运算。单击"启动 multiply 动作的 IntentService"按钮，再开启新线程，运行一次乘法运算。

1．打开基础工程

打开"基础工程"文件夹中的"C0604"工程，该工程已经包含 MainActivity 及布局的文件。

图 6-10　运行效果

2. 新建 IntentService

选择"/java/com/vt/c0604"文件夹,单击右键并选择【New】→【Service】→【Service (IntentService)】命令。在打开的"New Android Component"对话框中,设置"Class Name"为"MyIntentService",单击"Finish"按钮。

```
/java/com/vt/c0604/MyIntentService.java
08  public class MyIntentService extends IntentService {
09      public static final String ACTION_ADD = "com.vt.c0604.action.add";
10      public static final String ACTION_MULTPLY = "com.vt.c0604.action.multiply";
11      private static final String EXTRA_PARAM1 = "com.vt.c0604.extra.PARAM1";
12      private static final String EXTRA_PARAM2 = "com.vt.c0604.extra.PARAM2";
13      private static final String TAG = "MyIntentService";
14      public MyIntentService() {
15          super("MyIntentService");
16          Log.i(TAG, "线程" + Thread.currentThread().getId() + ":MyIntentService()构造方
    法被调用,MyIntentService 实例哈希码为"+this.hashCode());
17      }
18      //静态方法: 启动 add 动作的 MyIntentService
19      public static void startActionAdd(Context context, int param1, int param2) {
20          Intent intent = new Intent(context, MyIntentService.class);
21          intent.setAction(ACTION_ADD);
22          intent.putExtra(EXTRA_PARAM1, param1);
23          intent.putExtra(EXTRA_PARAM2, param2);
24          context.startService(intent);
25      }
26      //静态方法: 启动 multiply 动作的 MyIntentService
27      public static void startActionMultiply(Context context,int param1, int param2) {
28          Intent intent = new Intent(context, MyIntentService.class);
29          intent.setAction(ACTION_MULTPLY);
30          intent.putExtra(EXTRA_PARAM1, param1);
31          intent.putExtra(EXTRA_PARAM2, param2);
32          context.startService(intent);
33      }
34      //重写 onHandleIntent,根据动作调用不同的方法
35      @Override
36      protected void onHandleIntent(Intent intent) {
37          if (intent != null) {
38              final String action = intent.getAction();
39              if (ACTION_ADD.equals(action)) {
40                  Log.i(TAG, "线程" + Thread.currentThread().getId() + ":调用加法运算方法,
    MyIntentService 实例哈希码为"+this.hashCode());
41                  final int param1 = intent.getExtras().getInt(EXTRA_PARAM1);
42                  final int param2 = intent.getExtras().getInt(EXTRA_PARAM2);
43                  handleActionAdd(param1, param2);
44              } else if (ACTION_MULTPLY.equals(action)) {
45                  Log.i(TAG, "线程" + Thread.currentThread().getId() + ":调用乘法运算方法,
    MyIntentService 实例哈希码为"+this.hashCode());
46                  final int param1 = intent.getExtras().getInt(EXTRA_PARAM1);
47                  final int param2 = intent.getExtras().getInt(EXTRA_PARAM2);
```

```
48              handleActionMultiply(param1, param2);
49          }
50      }
51  }
52  //加法运算方法
53  public void handleActionAdd(int param1, int param2) {
54      Log.i(TAG, "线程" + Thread.currentThread().getId() + ":" + param1 + "+" + param2
    + "=" + (param1 + param2)+",MyIntentService 实例哈希码为"+this.hashCode());
55  }
56  //乘法运算方法
57  private void handleActionMultiply(int param1, int param2) {
58      Log.i(TAG, "线程" + Thread.currentThread().getId() + ":" + param1 + "*" + param2
    + "=" + (param1 * param2)+",MyIntentService 实例哈希码为"+this.hashCode());
59  }
60  }
```

第 09 ~ 12 行定义了 4 个字符串常量作为动作名称和参数名称,字符串的值不重复且合法即可。第 16 行 Thread.currentThread().getId()获取线程的 id 值,this.hashCode()获取实例的哈希码。第 20 ~ 24 行和第 28 ~ 32 行都通过 Intent 启动 MyIntentService。第 36 ~ 51 行重写 onHandleIntent(Intent intent)方法处理 Intent,根据不同的 action 调用不同的方法。第 38 行 intent.getAction()获取 Intent 的 action 值。第 41 ~ 42 行和第 46 ~ 47 行获取 Intent 传入的参数。

3. 主界面的 Activity

```
/java/com/vt/c0604/MainActivity.java
08  public class MainActivity extends AppCompatActivity {
09      private static final String TAG = "MyIntentService";
10      @Override
11      protected void onCreate(Bundle savedInstanceState) {
12          super.onCreate(savedInstanceState);
13          setContentView(R.layout.activity_main);
14          Log.i(TAG, "线程" + Thread.currentThread().getId() + ":MainActivity.onCreate()被调用");
15          //"启动 add 动作的 IntentService"按钮单击事件
16          findViewById(R.id.button_start_foo).setOnClickListener(new View.OnClickListener() {
17              @Override
18              public void onClick(View v) {
19                  MyIntentService.startActionAdd(MainActivity.this, 13, 7);
20              }
21          });
22          //"启动 multiply 动作的 IntentService"按钮单击事件
23          findViewById(R.id.button_start_baz).setOnClickListener(new View.OnClickListener() {
24              @Override
25              public void onClick(View v) {
26                  MyIntentService.startActionMultiply(MainActivity.this, 43, 27);
27              }
28          });
29      }
30  }
```

第 14 行输出主线程的 id 值。第 19 行和第 26 行直接使用类名调用 MyIntentService 的静态方法。

4．测试运行

运行后，依次单击"启动 add 动作的 IntentService"按钮和"启动 multiply 动作的 IntentService"按钮。观察"Logcat"窗口输出的结果（如图 6-11 所示），首先在 onCreate() 方法内输出主线程的 id。单击按钮后，通过类名直接调用静态方法，构造方法会被调用，构造方法运行时输出的线程 id 与主线程相同，说明在主线程中运行。onHandleIntent(Intent intent)内输出的线程 id 与主线程不同，内部调用的方法输出的线程 id 与其相同，说明已经自动启用了新线程，调用的方法也会在新线程中运行。

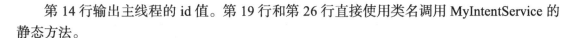

图 6-11　"Logcat"窗口输出的结果

6.3　广播接收器

6.3.1　广播接收器概述

在 Android 中，可以通过发送广播通知其他 App 所发生的事件，以便其他 App 进行相应处理。例如，电量低或充足、刚启动、插入耳机、输入法改变等发生时系统都会向所有 App 发送广播，这类广播称为系统广播（常用的系统广播如表 6-5 所示）。用户也可以自定义广播，发送给指定的广播接收器。

表 6-5　常用的系统广播

类　　型	说　　明
ConnectivityManager.CONNECTIVITY_ACTION	网络连接状态改变的广播
Intent.ACTION_AIRPLANE_MODE_CHANGED	开关飞行模式时发送该广播
Intent.ACTION_BATTERY_CHANGED	电池状态改变时发送该广播，不能静态注册使用
Intent.ACTION_BOOT_COMPLETED	系统启动完成后发送该广播
Intent.ACTION_CAMERA_BUTTON	"拍照"按钮被按下时发送该广播
Intent.ACTION_CONFIGURATION_CHANGED	系统配置（方向、区域设置等）改变时发送该广播
Intent.ACTION_DATE_CHANGED	修改日期后发送该广播
Intent.ACTION_DREAMING_STARTED	屏保开始后发送该广播
Intent.ACTION_DREAMING_STOPPED	屏保停止后发送该广播
Intent.ACTION_HEADSET_PLUG	耳机插孔上插拔耳机时发送该广播
Intent.ACTION_INPUT_METHOD_CHANGED	输入法改变时发送该广播
Intent.ACTION_LOCALE_CHANGED	系统设置的地区改变时发送该广播
Intent.ACTION_REBOOT	系统重新启动时发送该广播

续表

类　　型	说　　明
Intent.ACTION_SCREEN_OFF	息屏时发送该广播
Intent.ACTION_SCREEN_ON	屏幕激活时发送该广播
Intent.ACTION_TIMEZONE_CHANGED	系统设置的时区改变时发送该广播
Intent.ACTION_TIME_CHANGED	系统设置的时间改变时发送该广播
Intent.ACTION_TIME_TICK	系统每分钟发送一次该广播，不能静态注册使用
Intent.ACTION_USER_PRESENT	屏幕唤醒且屏保消失后发送该广播
Intent.ACTION_USER_UNLOCKED	开机解锁后发送该广播，不能静态注册使用

BroadcastReceiver（android.content.BroadcastReceiver）是处理系统和应用程序之间通信的组件（BroadcastReceiver 类的常用方法如表 6-6 所示）。常用 ContextWrapper.registerReceiver (BroadcastReceiver, IntentFilter)方法注册广播接收器，常用 ContextWrapper. unregisterReceiver (BroadcastReceiver)方法注销广播接收器。

表 6-6　BroadcastReceiver 类的常用方法

类型和修饰符	方　　法
	BroadcastReceiver () 构造方法
final void	abortBroadcast() 中止有序广播，仅适用于 sendOrderedBroadcast(Intent,String)发送的广播
final void	clearAbortBroadcast() 恢复有序广播
final boolean	getAbortBroadcast() 获取是否中止有序广播
final int	getResultCode() 获取前一个接收器设置的返回码
final String	getResultData() 获取前一个接收器设置的 String 型的返回码
final Bundle	getResultExtras(boolean makeMap) 获取前一个接收器设置的 bundle 型的返回码。makeMap 参数设置返回值是否可为空
final boolean	isOrderedBroadcast() 判断接收到的广播是否是有序广播
abstract void	onReceive(Context context, Intent intent) 接收到广播后调用该方法
final void	setResult(int code, String data, Bundle extras) 设置当前接收器广播的结果，仅适用于 sendOrderedBroadcast(Intent,String)方法发送的广播
final void	setResultCode(int code) 设置当前接收器广播的返回码，仅适用于 sendOrderedBroadcast(Intent,String)方法发送的广播
final void	setResultData(String data) 设置当前接收器广播的 String 型的返回码，仅适用于 sendOrderedBroadcast(Intent,String)方法发送的广播
final void	setResultExtras(Bundle extras) 设置当前接收器广播的 bundle 型的返回码，仅适用于 sendOrderedBroadcast(Intent,String)方法发送的广播

BroadcastReceiver 需要在 AndroidManifest.xml 文件中添加<receiver>标签后才能被调

用，<receiver>标签属性（如表6-7所示）大多无法通过BroadcastReceiver类的方法进行设置和获取。

表6-7 <receiver>标签属性

属　　性	说　　明
android:directBootAware	设置是否可以在用户解锁前直接启动
android:enabled	设置是否可以被实例化
android:exported	设置是否可以接收来自其他App的消息
android:label	设置标签名称
android:name	设置实现BroadcastReceiver的类名
android:permission	设置向BroadcastReceiver发送消息须具备的权限名称
android:process	设置运行BroadcastReceiver的进程名称

6.3.2　接收广播

接收广播有两种方式：显式接收和隐式接收。显式接收使用ContextWrapper.registerReceiver (BroadcastReceiver, IntentFilter)方法注册广播接收器，隐式接收在AndroidManifest.xml文件中使用<intent-filter>标签设置接收的广播类型。

广播消息本身会被封装在一个Intent对象中，该Intent对象的操作字符串会标识所发生的事件（如android.intent.action.AIRPLANE_MODE）。该Intent对象可能还包含绑定到其extra字段中的附加信息。例如，飞行模式Intent对象通过包含布尔值extra来设置是否已开启飞行模式。

> 提示：广播接收的限制
> 从Android 8.0（API level 26）开始，系统对AndroidManifest.xml文件声明的广播接收器进行了额外的限制，对于大多数的系统广播只能使用显式接收。

6.3.3　实例工程：显式和隐式接收广播

本实例演示了广播接收器使用显式和隐式接收广播（如图6-12所示）。单击"注册广播接收器"按钮，不但能够对网络、屏幕和电量等隐式广播进行接收，还能对自定义的显式广播进行接收。

1．打开基础工程

打开"基础工程"文件夹中的"C0605"工程，该工程已经包含MainActivity、Util类及布局的文件。Util类中包含getMobileType(NetworkInfo info)方法，用于获取移动网络的类型。

2．新建BroadcastReceiver

选择"/java/com/vt/c0605"文件夹，单击右键并选择【New】→【Other】→【Broadcast Receiver】命令。在打开的"New Android Component"对话框中，设置"Class Name"为"MyBroadcast- Receiver"，单击"Finish"按钮。

图6-12　运行效果

```
/java/com/vt/c0605/MyBroadcastReceiver.java
11  public class MyBroadcastReceiver extends BroadcastReceiver {
12      private static final String TAG = "MyBroadcastReceiver";
13      private int mBatteryLevel;//缓存电量值
14      @Override
15      public void onReceive(Context context, Intent intent) {
16          Log.d(TAG, "----------Receiver----------");
17          Log.d(TAG, "Action: " + intent.getAction());
18          Log.d(TAG, "URI: " + intent.toUri(Intent.URI_INTENT_SCHEME));
19          //网络
20          if (ConnectivityManager.CONNECTIVITY_ACTION.equals(intent.getAction())) {
21              //获取联网状态的NetworkInfo对象
22              NetworkInfo info = ((ConnectivityManager) context.getSystemService (Context.CONNECTIVITY_SERVICE)).getActiveNetworkInfo();
23              if (info != null && info.isAvailable()) {
24                  if ((info.getType() == ConnectivityManager.TYPE_WIFI)) {
25                      Log.d(TAG, "正在使用wifi网络");
26                  } else if (info.getType() == ConnectivityManager.TYPE_MOBILE) {
27                      String mobileType = Util.getMobileType(info);
28                      Log.d(TAG, "正在使用" + mobileType + "移动网络");
29                  }
30              } else {
31                  Log.d(TAG, "网络关闭");
32              }
33          }
34          //屏幕
35          if (Intent.ACTION_SCREEN_ON.equals(intent.getAction())) {
36              Log.i(TAG, "屏幕打开");
37          } else if (Intent.ACTION_SCREEN_OFF.equals(intent.getAction())) {
38              Log.i(TAG, "屏幕关闭");
39          } else if (Intent.ACTION_USER_PRESENT.equals(intent.getAction())) {
40              Log.i(TAG, "屏幕解锁");
41          }
42          //电量
43          if (Intent.ACTION_BATTERY_OKAY.equals(intent.getAction())) {
44              Log.i(TAG, "电量已满");
45          } else if (Intent.ACTION_BATTERY_LOW.equals(intent.getAction())) {
46              Log.i(TAG, "电量不足");
47          } else if (Intent.ACTION_BATTERY_CHANGED.equals(intent.getAction())) {
48              int level = intent.getIntExtra(BatteryManager.EXTRA_LEVEL, 0);
49              if (mBatteryLevel != level) {
50                  Log.i(TAG, "电量: " + level + "%");
51                  mBatteryLevel = level;
52              }
53          }
54          //自定义广播
55          if ("MyBroadcastReceiver.Custom".equals(intent.getAction())) {
56              Log.d(TAG, "自定义广播的info: " + intent.getStringExtra("info"));
57          }
58      }
59  }
```

第 15～58 行重写 onReceive(Context context, Intent intent)方法，接收并处理广播。第 17 行 intent.getAction()获取广播的类型。第 18 行 intent.toUri(Intent.URI_INTENT_SCHEME) 查看广播的完整信息。第 20 行 ConnectivityManager.CONNECTIVITY_ACTION.equals (intent.getAction())判断是否是网络连接状态改变的广播。第 24 和 26 行 info.getType() 获取网络类型，分别通过 ConnectivityManager.TYPE_WIFI 和 ConnectivityManager. TYPE_MOBILE 两个常量来判断所使用的网络类型。第 27 行 Util.getMobiletype()获取使用的无线通信技术类型。第 55 行判断接收的广播是否是 MyBroadcastReceiver.Custom 类型广播，该自定义类型通过 6.3.5 节的发送广播实例进行发送。第 56 行 intent.getStringExtra("info") 获取广播发送 info 字段的附加信息。

```
/manifests/AndroidManifest.xml
12    <receiver
13        android:name=".MyBroadcastReceiver"
14        android:enabled="true"
15        android:exported="true">
16        <intent-filter>
17            <action android:name="MyBroadcastReceiver.Custom" />
18        </intent-filter>
19    </receiver>
```

第 16～18 行<intent-filter>标签用于隐式接收广播，并设置接收广播的自定义类型为 MyBroadcast- Receiver.Custom。

3. 主界面的 Activity

```
/java/com/vt/c0605/MainActivity.java
10  public class MainActivity extends AppCompatActivity {
11      private MyBroadcastReceiver myReceiver = new MyBroadcastReceiver();
12      private boolean mRegistered = false;
13      @Override
14      protected void onCreate(Bundle savedInstanceState) {
15          super.onCreate(savedInstanceState);
16          setContentView(R.layout.activity_main);
17          //注册广播接收器
18          findViewById(R.id.button_register).setOnClickListener(new View.OnClickListener() {
19              @Override
20              public void onClick(View v) {
21                  IntentFilter intentFilter = new IntentFilter();
22                  intentFilter.addAction(ConnectivityManager.CONNECTIVITY_ACTION);
23                  intentFilter.addAction(Intent.ACTION_SCREEN_ON);
24                  intentFilter.addAction(Intent.ACTION_SCREEN_OFF);
25                  intentFilter.addAction(Intent.ACTION_USER_PRESENT);
26                  intentFilter.addAction(Intent.ACTION_BATTERY_OKAY);
27                  intentFilter.addAction(Intent.ACTION_BATTERY_LOW);
28                  intentFilter.addAction(Intent.ACTION_BATTERY_CHANGED);
29                  //intentFilter.addAction("com.vt.c0701.MyBroadcastReceiver.ss");
30                  registerReceiver(myReceiver, intentFilter);
31                  mRegistered = true;
```

```
32              }
33          });
34          //注销广播接收器
35          findViewById(R.id.button_unregister).setOnClickListener(new View.OnClickListener() {
36              @Override
37              public void onClick(View v) {
38                  if(mRegistered) {
39                      unregisterReceiver(myReceiver);
40                      mRegistered = false;
41                  }
42              }
43          });
44          //启动新窗口
45          findViewById(R.id.button_start_activity).setOnClickListener(new View.OnClickListener() {
46              @Override
47              public void onClick(View v) {
48                  Intent i = new Intent(MainActivity.this,MainActivity.class);
49                  startActivity(i);
50              }
51          });
52      }
53      @Override
54      protected void onDestroy() {
55          super.onDestroy();
56          if(mRegistered) {
57              unregisterReceiver(myReceiver);
58          }
59      }
60  }
```

第 11 行声明并实例化一个 MyBroadcastReceiver 类的 myReceiver 实例。第 12 行声明并初始化 mRegistered 记录是否已经注册 myReceiver。第 22 ~ 28 行添加了 7 种类型的系统广播。第 29 行使用 "//" 注释自定义类型的广播，当 AndroidManifest.xml 文件中没有设置隐式接收广播的自定义类型时，将 "//" 删除后就可以设置显式接收的自定义类型。第 30 行 registerReceiver(myReceiver, intentFilter)注册 myReceiver 及过滤器 intentFilter。第 38 ~ 41 行判断是否已经注册 myReceiver，如果注册过，则使用 unregisterReceiver(myReceiver)注销。这里添加判断的原因是如果未注册，则注销 myReceiver，App 会报错并直接退出。第 45 ~ 51 行启动一个新的 MainActivity 实例，验证每个 Activity 都可以注册一个 myReceiver 实例，多个 MainActivity 实例的 myReceiver 实例可以同时注册运行。第 54 ~ 59 行重写 onDestroy()方法，如果已经注册 myReceiver，则进行注销，因为在未注销时关闭 Activity 会报错并直接退出。

4. 测试运行

运行后，单击 "注册广播接收器" 按钮会在 "Logcat" 窗口中输出网络连接和电量的相关信息，再单击电源按钮后屏幕会熄灭，在 "Logcat" 窗口中输出屏幕关闭的相关信息（如图 6-13 所示）。单击 "启动新窗口" 按钮，启动新窗口，再单击新窗口中的 "注册广播接收器"按钮，会在"Logcat"窗口中输出两条广播的处理信息。后退关闭新窗口，会注销新窗口的广播接收器，此时在"Logcat"窗口中只会输出一条广播的处理信息。

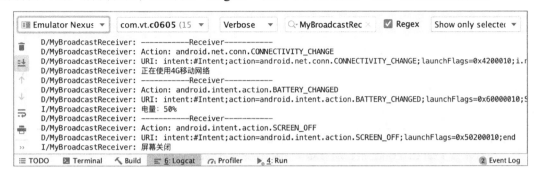

图 6-13 "Logcat"窗口输出的结果

6.3.4 发送广播

系统广播只能由 Android 系统发出，用户只能发送自定义广播。自定义广播主要分为标准广播和有序广播两种。标准广播是指广播会被所有的接收者接收到，不可以被拦截和被修改；常用 Context.sendBroadcast(Intent)方法发送标准广播。有序广播是指广播按照优先级逐级向下传递，接收者可以修改广播数据，也可以中止广播；常用 Context.sendOrderedBroadcast(Intent,String)方法发送有序广播。

6.3.5 实例工程：发送标准广播和有序广播

本实例演示了发送标准广播和有序广播的方法（如图 6-14 所示）。单击相应的按钮会在"Logcat"窗口中显示发送的相应信息。

1. 打开基础工程

打开"基础工程"文件夹中的"C0606"工程，该工程已经包含 MainActivity 及布局的文件。

2. 新建 BroadcastReceiver

在"/java/com/vt/c0606"文件夹中，新建两个自定义广播接收器类：MyBroadcastReceiver1 和 MyBroadcastReceiver2。

图 6-14 运行效果

```
/java/com/vt/c0606/MyBroadcastReceiver1.java
08    public class MyBroadcastReceivers1 extends BroadcastReceiver {
09        private static final String TAG = "MyBroadcastReceiver";
10        @Override
11        public void onReceive(Context context, Intent intent) {
12            Log.d(TAG, "-----------Receivers1-----------");
13            Log.d(TAG, "Action: " + intent.getAction());
```

```
14            Log.d(TAG, "URI: " + intent.toUri(Intent.URI_INTENT_SCHEME));
15            Log.d(TAG, "自定义广播的info: " + intent.getStringExtra("info"));
16        }
17    }
```

第 11~16 行重写 onReceive(Context context, Intent intent)方法，接收并处理广播。第 13 行 intent.getAction()获取广播类型。第 14 行 intent.toUri(Intent.URI_INTENT_SCHEME)接收广播的完整数据。第 15 行 intent.getStringExtra("info")获取 info 字段的额外数据。

```
/java/com/vt/c0606/MyBroadcastReceiver2.java
08  public class MyBroadcastReceivers2 extends BroadcastReceiver {
09      private static final String TAG = "MyBroadcastReceiver";
10      @Override
11      public void onReceive(Context context, Intent intent) {
12          Log.d(TAG, "-----------Receivers2-----------");
13          Log.d(TAG, "Action: " + intent.getAction());
14          Log.d(TAG, "URI: " + intent.toUri(Intent.URI_INTENT_SCHEME));
15          Log.d(TAG, "自定义广播的info: " + intent.getStringExtra("info"));
16          if(intent.getBooleanExtra("stop",false)){
17              abortBroadcast();
18          }
19      }
20  }
```

第 16 行 intent.getBooleanExtra("stop",false)获取 boolean 型 stop 字段的数据，如果无法获取到，则返回 false。第 17 行 abortBroadcast()中止有序广播继续传递。

```
/manifests/AndroidManifest.xml
11  <receiver
12      android:name=".MyBroadcastReceivers2"
13      android:enabled="true"
14      android:exported="true">
15      <intent-filter android:priority="200">
16          <action android:name="OrderedBroadcast.Custom" />
17      </intent-filter>
18  </receiver>
19  <receiver
20      android:name=".MyBroadcastReceivers1"
21      android:enabled="true"
22      android:exported="true">
23      <intent-filter android:priority="1">
24          <action android:name="OrderedBroadcast.Custom" />
25      </intent-filter>
26  </receiver>
```

第 15 行和第 23 行的 android:priority 属性设置有序广播的优先级，属性值范围是 [-1000,1000]。第 16 行和第 24 行设置接收广播的类型为 OrderedBroadcast.Custom，该类型属于自定义广播类型，合法的字符串通常都可以作为自定义广播类型的名称。

3. 主界面的 Activity

```
/java/com/vt/c0606/MainActivity.java
14   //发送标准广播
15   findViewById(R.id.button_send_broadcast).setOnClickListener(new View.OnClickListener() {
16       @Override
17       public void onClick(View v) {
18           Intent intent=new Intent();
19           intent.setAction("MyBroadcastReceiver.Custom");
20           intent.putExtra("info","悄悄地告诉你一个秘密^6^");
21           intent.addFlags(Intent.FLAG_ACTIVITY_PREVIOUS_IS_TOP);
22           sendBroadcast(intent);
23       }
24   });
25   //发送有序广播(连续传递)
26   findViewById(R.id.button_send_ordered_broadcast).setOnClickListener(new
     View.OnClickListener() {
27       @Override
28       public void onClick(View v) {
29           Intent intent=new Intent();
30           intent.setAction("OrderedBroadcast.Custom");
31           intent.putExtra("info","悄悄地告诉你一个秘密^6^");
32           intent.addFlags(Intent.FLAG_ACTIVITY_PREVIOUS_IS_TOP);
33           sendOrderedBroadcast(intent,null);
34       }
35   });
36   //发送有序广播(传递一次)
37   findViewById(R.id.button_send_ordered_broadcast_stop).setOnClickListener(new
     View.OnClickListener() {
38       @Override
39       public void onClick(View v) {
40           Intent intent=new Intent();
41           intent.setAction("OrderedBroadcast.Custom");
42           intent.putExtra("stop",true);
43           intent.putExtra("info","悄悄地告诉你一个秘密 不要告诉别人哦^6^");
44           intent.addFlags(Intent.FLAG_ACTIVITY_PREVIOUS_IS_TOP);
45           sendOrderedBroadcast(intent,null);
46       }
47   });
```

第 18～22 行发送的是标准广播，intent.setAction("MyBroadcastReceiver.Custom")设置发送广播的类型，intent.putExtra("info","悄悄地告诉你一个秘密^6^")设置 info 字段附加的字符串数据，intent.addFlags(Intent.FLAG_ACTIVITY_PREVIOUS_IS_TOP)设置 flag 标记，设置该标记是为了突破 Android 8.0 版本对隐式广播的限制。第 29～33 行和第 40～45 行发送的都是有序广播，接收到有序广播的对象能够决定是否继续将广播进行传递。

4. 测试运行

运行后，自上而下依次单击三个按钮，在"Logcat"窗口中观察输出的结果（如图 6-15 所示）。第一个按钮发送的广播，需要安装 6.3.3 节中的实例才能正确接收到。

图 6-15 "Logcat"窗口输出的结果

6.4 习 题

1. 通过 Service 模拟显示微信收到信息时状态栏显示的通知。
2. 通过 BroadcastReceiver 实现在耳机插孔上插拔耳机时提醒用户注意调整音量的提示。
3. 通过 BroadcastReceiver 实现两个 App 之间通过自定义广播进行数据通信。

第 7 章 数据存储与共享

App 通常需要保存用户的数据,有些保存在本地,有些则保存在服务器端。少量数据的本地存储可以使用共享偏好设置,大量数据的本地存储可以使用 SQLite。如果将数据保存在服务器端,通常使用 JSON 格式发送数据,然后通过服务器端的程序进行保存。App 之间单向数据传递适合使用广播,而双向数据传递适合使用内容提供者。

7.1 共享偏好设置

7.1.1 SharedPreferences 概述

SharedPreferences(android.content.SharedPreferences)是用于本地存储共享数据的轻量级接口,使用 XML 文件保存简单数据并存储在缓存文件夹中,特别适合本地存储用户的个人数据,敏感的个人数据最好加密后存储。在物理设备上可以直接通过文件浏览 App 查看该文件。在 Android Studio 中使用虚拟设备运行 App 时,需要通过 "Device File Explorer" 浏览虚拟设备的文件,在 "/data/data" 文件夹中可以找到对应 App 的文件夹,其内部的 "shared_prefs" 文件夹用于存储 SharedPreferences 创建的 XML 文件(如图 7-1 所示)。打开该 XML 文件,可以看到存储数据的标签。由于 SharedPreferences 存储的数据是未经过加密的,而且用户可以更改,所以对于重要数据应该进行加密处理或通过网络校验。

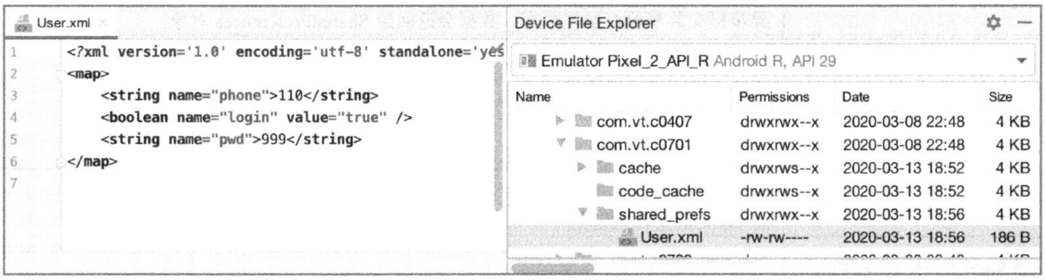

图 7-1 "Device File Explorer" 窗口

使用 Context.getSharedPreferences(String name, int mode)方法可以获取本地存储的共享偏好设置数据,其中 name 参数是存储共享偏好设置数据的文件名称,mode 参数多使用 Context. MODE_PRIVATE,其余 mode 参数已在 API level 23 中过期。SharedPreferences 接口主要用于获取共享偏好设置数据(如表 7-1 所示),SharedPreferences.Editor 接口主要用于写入、修改和清除共享偏好设置数据(如表 7-2 所示)。

表 7-1　SharedPreferences 接口的常用方法

类型和修饰符	方　　法
abstract boolean	contains(String key) 检查是否包含指定名称的共享偏好设置
Abstract SharedPreferences.Editor	edit() 为共享偏好设置创建一个编辑器，通过该编辑器可以对偏好设置数据进行修改，并自动提交返回给 SharedPreferences 对象
abstract Map<String, ?>	getAll() 获取所有的偏好设置数据
abstract boolean	getBoolean(String key, boolean defValue) 获取 boolean 型的键值数据，如果为空则返回 defValue 作为默认值
abstract float	getFloat(String key, float defValue) 获取 float 型的键值数据，如果为空则返回 defValue 作为默认值
abstract int	getInt(String key, int defValue) 获取 int 型的键值数据，如果为空则返回 defValue 作为默认值
abstract long	getLong(String key, long defValue) 获取 long 型的键值数据，如果为空则返回 defValue 作为默认值
abstract String	getString(String key, String defValue) 获取 String 型的键值数据，如果为空则返回 defValue 作为默认值
abstract Set<String>	getStringSet(String key, Set<String> defValue) 获取 Set<String>型的键值数据，如果为空则返回 defValue 作为默认值
abstract void	registerOnSharedPreferenceChangeListener(SharedPreferences.OnSharedPreferenceChangeListener listener) 注册监听共享偏好设置数据改变的监听器
abstract void	unregisterOnSharedPreferenceChangeListener(SharedPreferences.OnSharedPreferenceChangeListener listener) 注销监听共享偏好设置数据改变的监听器

表 7-2　SharedPreferences.Editor 接口的常用方法

类型和修饰符	方　　法
abstract void	apply() 异步提交共享偏好设置数据，数据会返回给 SharedPreferences 对象
abstract SharedPreferences.Editor	clear() 清除共享偏好设置数据
abstract boolean	commit() 提交共享偏好设置的操作，返回值为是否写入 XML 文件成功
abstract SharedPreferences.Editor	putBoolean(String key, boolean value) 将 boolean 值存入键值，在 commit()或 apply()方法执行后写入 XML 文件
abstract SharedPreferences.Editor	putFloat(String key, float value) 将 float 值存入键值，在 commit()或 apply()方法执行后写入 XML 文件
abstract SharedPreferences.Editor	putLong(String key, long value) 将 long 值存入键值，在 commit()或 apply()方法执行后写入 XML 文件
abstract SharedPreferences.Editor	putString(String key, String value) 将 String 值存入键值，在 commit()或 apply()方法执行后写入 XML 文件
abstract SharedPreferences.Editor	putStringSet(String key, Set<String> value) 将 Set<String>值存入键值，在 commit()或 apply()方法执行后写入 XML 文件
abstract SharedPreferences.Editor	remove(String key) 在编辑器中标记应删除的键值，在 commit()方法执行后删除该键值

7.1.2 实例工程：用户登录

本实例演示了 App 常用的登录功能（如图 7-2 所示），登录后保存用户的登录数据，再次打开 App 时会自动登录。登录后单击"退出"按钮，可以清除用户登录数据，再次打开 App 时需要重新登录。

图 7-2　运行效果

1．打开基础工程

打开"基础工程"文件夹中的"C0701"工程，该工程已经包含 MainActivity、SubActivity 及布局的文件。

2．主界面的 Activity

```
/java/com/vt/c0701/MainActivity.java
12  public class MainActivity extends AppCompatActivity {
13      private Context mContext;
14      @Override
15      protected void onCreate(Bundle savedInstanceState) {
16          super.onCreate(savedInstanceState);
17          setContentView(R.layout.activity_main);
18          mContext = this;
19          checkLogin();
20          //登录按钮单击事件
21          findViewById(R.id.btn_login).setOnClickListener(new View.OnClickListener() {
22              @Override
23              public void onClick(View v) {
24                  String phone = ((EditText) findViewById(R.id.edit_text_phone)).getText().toString().trim();
```

```
25              String pwd = ((EditText) findViewById(R.id.edit_text_pwd)).getText().
    toString().trim();
26              if (phone.equals("110") && pwd.equals("999")) {
27                  SharedPreferences sp = mContext.getSharedPreferences("User",
    Context.MODE_PRIVATE);
28                  SharedPreferences.Editor editorSP = sp.edit();
29                  editorSP.putBoolean("login", true);
30                  editorSP.putString("phone", phone);
31                  editorSP.putString("pwd", pwd);
32                  editorSP.commit();
33                  Intent intent = new Intent(mContext, SubActivity.class);
34                  startActivity(intent);
35              } else {
36                  ((TextView) findViewById(R.id.text_view_error)).setText("密码错误！");
37              }
38          }
39      });
40  }
41  //自动登录
42  private void checkLogin(){
43      SharedPreferences sp = mContext.getSharedPreferences("User", Context.MODE_PRIVATE);
44      if(sp.getBoolean("login", false)){
45          Intent intent = new Intent(mContext, SubActivity.class);
46          startActivity(intent);
47      }
48  }
49 }
```

第 24、25 行 trim()方法删除字符串两侧的空字符。第 26 行使用固定的字符串判断登录是否成功，实际项目中会将 phone 和 pwd 加密后发送到服务器端进行判断并返回结果。第 27 行 mContext.getSharedPreferences("User", Context.MODE_PRIVATE)获取 User.xml 文件中的共享偏好设置数据，如果 User.xml 文件不存在，则在 commit()方法执行后创建该文件。第 28 行 sp.edit()为共享偏好设置创建一个编辑器。第 29～31 行放入准备向共享偏好设置写入的键值数据。第 32 行 editorSP.commit()方法提交对共享偏好设置的操作。第 42～48 行是自动登录的方法，通过 SharedPreferences 存储的 login 键值判断是否已经登录；如果已经登录，则直接启动 SubActivity。

3. 登录后界面的 Activity

```
/java/com/vt/c0701/SubActivity.java
10  public class SubActivity extends AppCompatActivity {
11      private Context mContext;
12      @Override
13      protected void onCreate(Bundle savedInstanceState) {
14          super.onCreate(savedInstanceState);
15          setContentView(R.layout.activity_sub);
16          mContext = this;
```

```
17         //获取登录用户数据
18         SharedPreferences sp = mContext.getSharedPreferences("User", Context.MODE_PRIVATE);
19         ((TextView)findViewById(R.id.text_view)).setText("欢迎"+sp.getString("phone",
    "")+"访问！");
20         //退出登录
21         findViewById(R.id.btn_quit).setOnClickListener(new View.OnClickListener() {
22             @Override
23             public void onClick(View v) {
24                 SharedPreferences sp = mContext.getSharedPreferences("User", Context.MODE_PRIVATE);
25                 SharedPreferences.Editor editorSP = sp.edit();
26                 editorSP.clear();
27                 editorSP.commit();
28                 finish();
29             }
30         });
31     }
32 }
```

第 18 行 mContext.getSharedPreferences("User", Context.MODE_PRIVATE)获取 User.xml 文件中的共享偏好设置数据。第 19 行通过 sp.getString("phone", "")获取存储的登录用户手机号码。第 26 行 editorSP.clear()清除共享偏好设置的所有数据。第 27 行 editorSP.commit() 提交对共享偏好设置的操作。

7.2 轻量级数据库

7.2.1 SQLite 概述

SQLite 是一个轻量级的关系型数据库，无服务器端，零配置，运算速度快，占用资源少，支持大部分标准 SQL 语法。每个字段所使用的存储类型都是动态的，共有 5 种存储类型：NULL、INTEGER、TEXT、REAL 和 BLOB，根据数据类型使用相应的存储类型进行存储（如表 7-3 所示）。SQLite 主要通过 SQLiteOpenHelper 类、SQLiteDatabase 类和 Cursor 类对其进行操作。

表 7-3 数据类型使用的存储类型规则

数 据 类 型	SQLite 使用的存储类型
NULL	NULL：存储 NULL 值
INT	INTEGER：存储有符号整数
INTEGER	
TINYINT	
SMALLINT	
MEDIUMINT	
BIGINT	
UNSIGNED BIG INT	
INT2	
INT8	

续表

数据类型	SQLite 使用的存储类型
CHARACTER(20)	TEXT：存储字符串类型数据
VARCHAR(255)	
VARYING CHARACTER(255)	
NCHAR(55)	
NATIVE CHARACTER(70)	
NVARCHAR(100)	
TEXT	
CLOB	
REAL	REAL：存储浮点数
DOUBLE	
DOUBLE PRECISION	
FLOAT	
BLOB	BLOB：存储二进制大数据
NUMERIC	如果转换后完全可逆，则以 INTEGER 或 REAL 类型存储；如果转换后不可逆，则以 TEXT 类型存储
DECIMAL(10,5)	
BOOLEAN	
DATE	
DATETIME	

SQLiteOpenHelper 类（android.database.sqlite.SQLiteOpenHelper）主要用于 SQLite 数据库的创建和版本管理（如表 7-4 所示），其子类需要实现 onCreate(SQLiteDatabase)方法和 onUpgrade(SQLiteDatabase, int, int)方法。如果数据库不存在，会自动创建扩展名为 db 的数据库文件。在"/data/data"文件夹中可以找到对应 App 的文件夹，其内部的"databases"文件夹用于存储 SQLite 数据库文件（如图 7-3 所示）。

表 7-4 SQLiteOpenHelper 类的常用方法

类型和修饰符	方法
	SQLiteOpenHelper(Context context, String name, SQLiteDatabase.CursorFactory factory, int version) 构造方法
void	close() 关闭数据库
String	getDatabaseName() 获取数据库名称
SQLiteDatabase	getReadableDatabase() 以只读方式获取数据库对象
SQLiteDatabase	getWritableDatabase() 获取数据库对象，具有读写权限
abstract void	onCreate(SQLiteDatabase db) 创建数据库后调用该方法
void	onDowngrade(SQLiteDatabase db, int oldVersion, int newVersion) 数据库版本降级时调用该方法
void	onOpen(SQLiteDatabase db) 数据库打开后调用该方法
abstract void	onUpgrade(SQLiteDatabase db, int oldVersion, int newVersion) 数据库版本升级时调用该方法

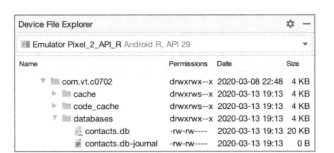

图 7-3 "Device File Explorer" 窗口

SQLiteDatabase 类（android.database.sqlite.SQLiteDatabase）用于对 SQLite 数据库进行操作（其常用方法如表 7-5 所示），可以使用 SQL 语句，支持事务管理。

表 7-5 SQLiteDatabase 类的常用方法

类型和修饰符	方法
long	insert(String table, String nullColumnHack, ContentValues values) 插入数据的便捷方法，返回值为插入数据所在行的索引号，插入失败时返回-1
int	delete(String table, String whereClause, String[] whereArgs) 删除数据的便捷方法，返回值为删除的行数
Cursor	query(String table, String[] columns, String selection, String[] selectionArgs, String groupBy, String having, String orderBy) 查询数据的便捷方法，返回值为数据集的指针
int	update(String table, ContentValues values, String whereClause, String[] whereArgs) 更新数据的便捷方法，返回值为更新的行数
void	beginTransaction() 以独占模式开始事务。事务可以嵌套，当外部事务结束时，该事务中完成的所有工作和所有嵌套事务都将提交或回滚。如果在未调用 setTransactionSuccessful()方法的情况下结束事务，则进行回滚
void	beginTransactionNonExclusive() 以立即模式开始事务
void	endTransaction() 结束事务
void	execSQL(String sql) 执行一个非 SELECT 或没有返回值的 SQL 语句
void	execSQL(String sql, Object[] bindArgs) 执行一个不是 SELECT、INSERT、UPDATE 或 DELETE 的 SQL 语句
String	getPath() 获取数据库文件的路径
int	getVersion() 获取数据库的版本
boolean	inTransaction() 当前线程是否有挂起的事务
boolean	isOpen() 判断数据库是否已经打开
boolean	isReadOnly() 判断数据库是否只读
boolean	needUpgrade(int newVersion) 判断 newVersion 是否大于当前版本

183

续表

类型和修饰符	方　　法
Cursor	rawQuery(String sql, String[] selectionArgs) 执行 SQL 语句并返回结果数据集
static int	releaseMemory() 尝试释放缓存
void	setTransactionSuccessful() 将当前事务标记为成功。在调用该方法和调用 endTransaction()方法之间不可对数据库进行任何操作，遇到任何错误则提交事务
void	setVersion(int version) 设置数据库版本

Cursor 接口（android.database.Cursor）用于对数据库查询的数据集进行读取（其常用方法如表 7-6 所示），通过移动指针指向不同的数据。

表 7-6　Cursor 接口的常用方法

类型和修饰符	方　　法	说　　明
abstract void	close()	关闭指针，并释放资源
abstract byte[]	getBlob(int columnIndex)	以字节数组的形式返回请求列的值
abstract int	getColumnCount()	获取列数和
abstract int	getColumnIndex(String columnName)	通过列的字段名获取列的索引号
abstract String[]	getColumnNames()	获取结果集中列的字段名称
abstract int	getCount()	获取结果集中的行数
abstract double	getDouble(int columnIndex)	根据列索引号获取 double 型的列值
abstract float	getFloat(int columnIndex)	根据列索引号获取 float 型的列值
abstract int	getInt(int columnIndex)	根据列索引号获取 int 型的列值
abstract long	getLong(int columnIndex)	根据列索引号获取 long 型的列值
abstract int	getPosition()	获取指针的位置
abstract short	getShort(int columnIndex)	根据列索引号获取 short 型的列值
abstract String	getString(int columnIndex)	根据列索引号获取 String 型的列值
abstract boolean	isAfterLast()	判断指针是否指向最后一行之后的位置
abstract boolean	isBeforeFirst()	判断指针是否指向第一行之前的位置
abstract boolean	isClosed()	判断指针是否关闭
abstract boolean	isFirst()	判断指针是否指向第一行
abstract boolean	isLast()	判断指针是否指向最后一行
abstract boolean	move(int offset)	指针从当前位置以指定的偏移量进行移动
abstract boolean	moveToFirst()	将指针移动到第一行
abstract boolean	moveToLast()	将指针移动到最后一行
abstract boolean	moveToNext()	将指针向下移动一行
abstract boolean	moveToPosition(int position)	将指针移动到指定的位置
abstract boolean	moveToPrevious()	将指针向上移动一行

7.2.2　实例工程：自定义通讯录

本实例演示了使用 SQLite 建立存储数据的通讯录，顶部使用 ListView 控件显示通讯录中的联系人，单击后在底部的 EditText 控件中显示该联系人的信息，可以通过底部的按钮对该联系人进行相应的操作或显示所有联系人（如图 7-4 所示）。

图 7-4 运行效果

1．打开基础工程

打开"基础工程"文件夹中的"C0702"工程，该工程已经包含 MainActivity、ContactListAdapter、ContactModel 及布局的文件。ContactListAdapter 类是显示联系人的 ListView 控件的适配器类，ContactModel 类是联系人的模型类。

2．ContactSQLiteOpenHelper 类

新建 ContactSQLiteOpenHelper 类，继承自 SQLiteOpenHelper 类，用于创建和更新 SQLite 数据库。

```
/java/com/vt/c0702/ContactSQLiteOpenHelper.java
07  public class ContactSQLiteOpenHelper extends SQLiteOpenHelper {
08      public ContactSQLiteOpenHelper(Context context, String name, SQLiteDatabase.CursorFactory factory, int version) {
09          super(context, name, factory, version);
10      }
11      @Override
12      public void onCreate(SQLiteDatabase db) {
13          //创建 contacts 表
14          db.execSQL("CREATE TABLE contacts(id INTEGER PRIMARY KEY AUTOINCREMENT,name VARCHAR(16),phone VARCHAR(11))");
15      }
16      @Override
17      public void onUpgrade(SQLiteDatabase db, int oldVersion, int newVersion) {
18          if (newVersion > oldVersion) {
19              //contacts 表添加 notes 列
20              db.execSQL("ALTER TABLE contacts ADD notes VARCHAR(20)");
21          }
22      }
23  }
```

第 08~10 行是 ContactSQLiteOpenHelper 的构造方法，super(context, name, factory, version)调用父类的构造方法。第 12~15 行重写 onCreate(SQLiteDatabase)方法，db.execSQL ("CREATE TABLE contacts(id INTEGER PRIMARY KEY AUTOINCREMENT,name VARCHAR(16), phone VARCHAR(11))")创建包含 id、name 和 phone 字段的数据表，通过 PRIMARY KEY AUTOINCREMENT 指定 id 字段为自动增长的主键。第 17~22 行重写 onUpgrade(SQLiteDatabase, int, int)方法，当数据库版本大于旧版本时，db.execSQL("ALTER TABLE contacts ADD notes VARCHAR(20)")为 contacts 表添加 notes 字段，备注联系人的相关信息。

3. 主界面的 Activity

```
/java/com/vt/c0702/MainActivity.java
46    //初始化
47    private void init(){
48        mContactSQLiteOpenHelper = new ContactSQLiteOpenHelper(mContext, mDataBaseName, null, 1);
49        mSQLiteDataBase = mContactSQLiteOpenHelper.getWritableDatabase();
50        mContactListAdapter = new ContactListAdapter(this, mContactModel);
51        mContactListView.setAdapter(mContactListAdapter);
52        listAllContacts();
53    }
54    //显示所有联系人
55    private void listAllContacts() {
56        mContactModel.clear();
57        Cursor cursor = mSQLiteDataBase.rawQuery("SELECT * FROM " + mTableName + " order by id desc", null);
58        while (cursor.moveToNext()) {
59            String id = cursor.getString(cursor.getColumnIndex("id"));
60            String name = cursor.getString(cursor.getColumnIndex("name"));
61            String phone = cursor.getString(cursor.getColumnIndex("phone"));
62            mContactModel.add(new ContactModel(id, name, phone));
63        }
64        cursor.close();
65        //更新 ListView
66        mContactListAdapter.notifyDataSetChanged();
67    }
68    //单击事件
69    @Override
70    public void onClick(View v) {
71        String id = mIdEditText.getText().toString();
72        String name = mNameEditText.getText().toString();
73        String phone = mPhoneEditText.getText().toString();
74        switch (v.getId()) {
75            case R.id.btn_list://显示所有联系人
76                listAllContacts();
77                break;
78            case R.id.btn_insert://添加联系人
79                mSQLiteDataBase.execSQL("INSERT INTO " + mTableName + "(name,phone) values(?,?)", new String[]{name, phone});
80                listAllContacts();
```

```
81              break;
82          case R.id.btn_query://查询联系人
83              mContactModel.clear();
84              Cursor cursor = mSQLiteDataBase.rawQuery("SELECT * FROM " + mTableName + "
    WHERE id = ? or name = ? or phone = ?", new String[]{id, name, phone});
85              while (cursor.moveToNext()) {
86                  String idQuery = cursor.getString(cursor.getColumnIndex("id"));
87                  String nameQuery = cursor.getString(cursor.getColumnIndex("name"));
88                  String phoneQuery = cursor.getString(cursor.getColumnIndex("phone"));
89                  mContactModel.add(new ContactModel(idQuery, nameQuery, phoneQuery));
90              }
91              cursor.close();
92              mContactListAdapter.notifyDataSetChanged();
93              break;
94          case R.id.btn_update://更新联系人
95              mSQLiteDataBase.execSQL("UPDATE " + mTableName + " SET name = ?,phone = ?
    WHERE id = ?", new String[]{name, phone, id});
96              listAllContacts();
97              break;
98          case R.id.btn_delete://删除联系人
99              mSQLiteDataBase.execSQL("DELETE FROM " + mTableName + " WHERE id = ?", new
    String[]{id});
100             listAllContacts();
101             break;
102         }
103     }
```

第 57 行创建以 id 字段降序排列查询 contacts 表内所有联系人的 cursor 对象。第 58 行 cursor.moveToNext()向下移动指针,当指针指向的位置存在数据时返回 true,否则返回 false。第 59～61 行 cursor.getColumnIndex()方法根据字段名称获取字段的索引号,cursor.getString()方法根据字段的索引号获取字符型数据。第 64 行 cursor.close()关闭指针,释放缓存数据。第 79 行"INSERT INTO " + mTableName + "(name,phone) values(?,?)"合成插入数据的 SQL 语句,两个?使用 name 和 phone 的变量值替换,但表名称不能使用?方式进行替换。第 80 行在插入数据后调用 listAllContacts()方法将新添加的联系人显示出来。第 84 行根据 id、name 或 phone 字段查询联系人,只要有一个字段相等即可。第 95 行根据 id 更新 name 和 phone 字段的数据。第 99 行删除指定 id 的联系人数据。

7.3 内容提供者

7.3.1 ContentProvider 概述

1. URI

URI(Universal Resource Identifier)是指通用资源标志符,Android 中的每种资源都可以用 URI 来表示。URI 有三种格式,如表 7-7 所示。

表 7-7 URI 的格式

格 式 规 则	格 式 样 例
scheme://authority[/path]	content://com.vt.contacts/people/47 content://media/external
scheme://host[:port][/path]	http://www.weiju2014.com http://www.meachol.com:80/weijian
scheme:scheme-specific-part	tel:13588888888 mailto:baizhe_22@qq.com

URI 作为 ContentProvider 授权的依据，还具有派生关系。例如，下面的三个 URI 中，第一个 URI 的 contact 表示 contact 数据表；第二个 URI 与第一个 URI 属于同一级别，#是一个通配符，表示数据表中的行号；第三个 URI 是前两个 URI 的派生 URI，类似于子类。

- content://com.vt.contacts/contact
- content://com.vt.contacts/contact/#
- content://com.vt.contacts/contact/friend

2. ContentProvider 类

ContentProvider 类（android.content.ContentProvider）用于提供进程之间安全地访问和修改数据的标准接口（其方法如表 7-8 所示），适用于为其他 App 提供复杂的数据或文件，向 Widget 公开应用数据，以及使用搜索框架提供自定义搜索建议。

虽然 App 内部也可以使用 ContentProvider 访问和修改数据，但不推荐这样使用，APP 内部直接访问或修改数据效率更高。ContentProvider 可以精细控制数据访问权限。也可以选择仅在 APP 内限制对 ContentProvider 的访问，授予访问其他应用数据的权限，或配置读取和写入数据的不同权限。

表 7-8 ContentProvider 类的方法

类型和修饰符	方 法
	ContentProvider() 构造方法
abstract boolean	onCreate() 创建时调用该方法
abstract Uri	insert(Uri uri, ContentValues values) 处理插入数据请求的方法，能够防止 SQL 注入式攻击
abstract int	delete(Uri uri, String selection, String[] selectionArgs) 处理删除数据请求的方法，能够防止 SQL 注入式攻击
abstract int	update(Uri uri, ContentValues values, String selection, String[] selectionArgs) 处理更新数据请求的方法，能够防止 SQL 注入式攻击
abstract Cursor	query(Uri uri, String[] projection, String selection, String[] selectionArgs, String sortOrder) 处理查询数据请求的方法，能够防止 SQL 注入式攻击
abstract String	getType(Uri uri) 处理通过 URI 获取 MIME 类型的方法
final String	getCallingPackage () 获取发送访问请求的包名称。如果当前没有访问请求，则返回 null
final Context	getContext() 获取运行 ContentProvider 的 Context 对象

ContentProvider 需要在 AndroidManifest.xml 文件中添加<provider>标签后才能被调用，<provider>标签属性（如表 7-9 所示）大多无法通过 ContentProvider 类的方法进行设置和获取。

表 7-9 <provider>标签属性

属　　性	说　　明
android:authorities	设置用于标识 ContentProvider 提供数据的 URI，多个 URI 用 ";" 分隔
android:directBootAware	设置是否可以在用户解锁前直接启动
android:name	设置实现 ContentProvider 的类名
android:enabled	设置是否可以实例化 ContentProvider
android:exported	设置是否可供其他 App 使用
android:initOrder	设置启动顺序优先级，数值越高，优先级越高
android:multiprocess	设置是否为每个进程提供一个 ContentProvider 对象
android:process	设置运行 ContentProvider 的进程名称
android:syncable	设置是否与服务器上的数据同步

3．ContentResolver 类

访问 ContentProvider 实现数据访问和修改，需要通过 ContentResolver 或 Intent 对象。ContentResolver 类（android.content.ContentResolver）用于对 URI 进行解析，然后将请求发送到对应的 ContentProvider。ContentResolver 类的方法如表 7-10 所示。

表 7-10 ContentResolver 类的方法

类型和修饰符	方　　法
ContentProviderResult[]	applyBatch(String authority, ArrayList<ContentProviderOperation> operations) 批量执行 ContentProviderOperation 对象，并返回其结果的数组
final int	delete(Uri url, String where, String[] selectionArgs) 发送删除数据请求的方法
final Uri	insert(Uri url, ContentValues values) 发送插入数据请求的方法
final int	update(Uri uri, ContentValues values, String where, String[] selectionArgs) 发送更新数据请求的方法
final Cursor	query(Uri uri, String[] projection, String selection, String[] selectionArgs, String sortOrder) 发送查询数据请求的方法
void	notifyChange(Uri uri, ContentObserver observer) 通知观察者数据更新的方法
final void	registerContentObserver(Uri uri, boolean notifyForDescendants, ContentObserver observer) 注册观察者类对象的方法，该方法在调用 notifyChange()方法后回调
final void	unregisterContentObserver(ContentObserver observer) 注销观察者类对象的方法

4．ContentObserver 类

ContentObserver 类（android.database.ContentObserver）用于接收 ContentResolver 对象和自身发送的 URI 更新通知（其方法如表 7-11 所示），适合在 ContentProvider 所在的 App 内使用，以便及时处理通过 ContentProvider 插入、更新或删除的数据。

表 7-11 ContentObserver 类的方法

类型和修饰符	方　　法
	ContentObserver(Handler handler) 构造方法
final void	dispatchChange(boolean selfChange, Uri uri) 向观察者发送更改通知。如果构造方法提供了处理程序，则对 onChange(boolean)方法的调用将被传递到处理程序的消息队列中；否则，立即调用 onChange(boolean)方法
void	onChange(boolean selfChange) 当内容发生更改时调用此方法，子类应重写此方法以处理内容更改
void	onChange(boolean selfChange, Uri uri) 当内容发生更改时调用此方法，子类应重写此方法以处理内容更改。为了确保在未提供 uri 参数的旧版本框架上正确操作，应实现 onChange(boolean)和 onChange(boolean,android.net.Uri)的重写

5．提供数据的流程

ContentProvider 不但可以向所在的 App 提供数据，还可以向其他 App 提供数据（其流程如图 7-5 所示）。请求数据的 App 使用 URI 向提供数据的 App 请求数据，提供数据的 App 通过 ContentProvider 访问数据库获取数据。当请求数据的 App 要求改变数据库中的数据成功时，可以通过 ContentObserver 获取数据的改变信息，然后进行 UI 界面中的数据更新或其他操作。

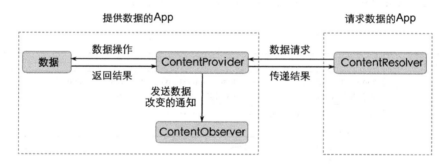

图 7-5　ContentProvider 提供数据的流程

7.3.2　实例工程：自定义内容提供者

本实例演示了两个 App 之间的数据访问和修改，模拟了系统通讯录提供内容服务的方式（如图 7-6 所示）。C0703 工程使用 SQLite 存储联系人的数据，使用自定义的 ContentProvider 提供联系人的查询、插入、更新和删除功能。C0704 工程并没有存联系人的数据，而是通过 ContentResolver 访问 C0703 工程提供的联系人数据的访问和修改功能。C0703 可以被看成服务器端，C0704 可以被看成客户端。

1．打开内容提供者的基础工程

打开"基础工程"文件夹中的"C0703"工程，该工程已经包含 MainActivity、ContactListAdapter、ContactModel 及布局的文件。ContactListAdapter 是显示联系人的 ListView 控件的适配器类，ContactModel 是联系人的模型类。

图 7-6　运行效果

2．内容提供者的 ContactContentProvider 类

选择"/java/com/vt/c0703"文件夹，单击右键并选择【new】→【Other】→【Content Provider】命令。在打开的"New Android Component"对话框（如图 7-7 所示）中，设置"Class Name"为"ContactContentProvider"，"URI Authorities"为"com.vt.contact"，单击"Finish"按钮完成 ContactContentProvider 类的创建。

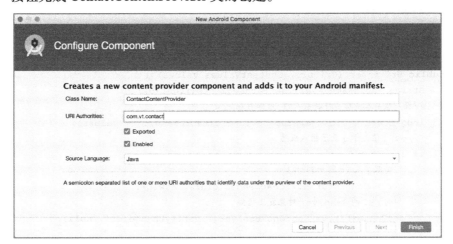

图 7-7　"New Android Component"对话框

/manifests/AndroidManifest.xml	
11	`<provider`
12	`    android:name=".ContactContentProvider"`
13	`    android:authorities="com.vt.contact"`
14	`    android:enabled="true"`
15	`    android:exported="true"></provider>`

第 11 ~ 15 行是新建 ContactContentProvider 类时自动生成的标签代码。第 13 行 android:authorities="com.vt.contact"表示能够接收包含"com.vt.contact"授权的 URI。

```
/java/com/vt/c0703/ContactContentProvider.java
12   public class ContactContentProvider extends ContentProvider {
13       private static final String AUTHORITIES = "com.vt.contact";
14       private static final UriMatcher URI_MATCHER;
15       private static final int CONTACTS_TABLE = 1;
16       private static final int CONTACTS_TABLE_AND_ID = 2;
17       private static ContactDBOpenHelper mContactDBOpenHelper;
18       private static SQLiteDatabase mSQLiteDatabase;
19       private static String mDataBaseName = "contacts.db";
20       //static 代码块,只会在类加载时执行一次,优化系统性能
21       static {
22           URI_MATCHER = new UriMatcher(UriMatcher.NO_MATCH);
23           URI_MATCHER.addURI(AUTHORITIES, "contacts", CONTACTS_TABLE);
24           URI_MATCHER.addURI(AUTHORITIES, "contacts/#", CONTACTS_TABLE_AND_ID);
25       }
26       //创建时调用
27       @Override
28       public boolean onCreate() {
29           //实例化 ContactDBOpenHelper
30           mContactDBOpenHelper = new ContactDBOpenHelper(this.getContext(), mDataBaseName,
     null, 1);
31           //获取具有写入权限的 SQLite 数据库
32           mSQLiteDatabase = mContactDBOpenHelper.getWritableDatabase();
33           return true;
34       }
35       //插入
36       @Override
37       public Uri insert(Uri uri, ContentValues values) {
38           String tableName = uri.getPathSegments().get(0);
39           //插入联系人成功后将联系人的 id 保存在 rowID 中
40           long rowID = mSQLiteDatabase.insert(tableName, "", values);
41           //rowID 值大于 0 表示插入成功
42           if (rowID > 0) {
43               //将联系人 id 附加到 URI
44               uri = ContentUris.withAppendedId(uri, rowID);
45               //向 URI 的 Resolver 对象发送通知
46               getContext().getContentResolver().notifyChange(uri, null);
47           }
48           return uri;
49       }
50       //更新
51       @Override
52       public int update(Uri uri, ContentValues values, String selection, String[]
     selectionArgs) {
53           int count = 0;
54           //获取 URI 中附加的联系人 id
55           String id = uri.getPathSegments().get(1);
```

```
56              //更新数据后将更新的行数赋值给count
57              count = mSQLiteDatabase.update(uri.getPathSegments().get(0), values, "id= " +
    id + (!TextUtils.isEmpty(selection) ? " AND (" + selection + ')' : ""), selectionArgs);
58              //向URI的ContentResolver对象发送改变的通知
59              getContext().getContentResolver().notifyChange(uri, null);
60              return count;
61          }
62          //删除
63          @Override
64          public int delete(Uri uri, String selection, String[] selectionArgs) {
65              int count = 0;
66              String tableName = uri.getPathSegments().get(0);
67              //判断URI匹配的常量值
68              switch (URI_MATCHER.match(uri)) {
69                  case CONTACTS_TABLE:
70                      //根据name和phone判断删除联系人并将删除的联系人数据行数赋值给count
71                      count = mSQLiteDatabase.delete(tableName, selection, selectionArgs);
72                      break;
73                  case CONTACTS_TABLE_AND_ID:
74                      //获取要删除的联系人id
75                      String id = uri.getPathSegments().get(1);
76                      //根据id删除联系人
77                      count = mSQLiteDatabase.delete(tableName, "id = " + id , selectionArgs);
78                      break;
79                  default:
80                      throw new IllegalArgumentException("Unknown URI " + uri);
81              }
82              getContext().getContentResolver().notifyChange(uri, null);
83              return count;
84          }
85          //查询
86          @Override
87          public Cursor query(Uri uri, String[] columns, String selection, String[]
    selectionArgs, String sortOrder) {
88              //获取联系人的id
89              String tableName = uri.getPathSegments().get(0);
90              //将查询结果的指针赋值给cursor
91              Cursor cursor = mSQLiteDatabase.query(tableName, columns, selection,
    selectionArgs, null, null, sortOrder);
92              return cursor;
93          }
94          //通过URI获取MIME类型
95          @Override
96          public String getType(Uri uri) {
97              int match = URI_MATCHER.match(uri);
98              switch (match) {
99                  case CONTACTS_TABLE:
100                     //vnd.android.cursor.dir表示数据集合
101                     return "vnd.android.cursor.dir/contacts";
102                 case CONTACTS_TABLE_AND_ID:
```

```
103                //vnd.android.cursor.item表示一组数据
104                return "vnd.android.cursor.item/contacts";
105            default:
106                return null;
107        }
108    }
109 }
```

第 21~25 行使用常量匹配 URI，content://com.vt.contact/contacts 与 CONTACTS_TABLE 相匹配，content://com.vt.contact/contacts/# 与 CONTACTS_TABLE_AND_ID 中相匹配。第 37~49 行重写 insert(Uri uri, ContentValues values)方法，使用 mSQLiteDatabase.insert (tableName, "", values)插入数据，如果插入数据成功，会将数据的行号保存在 rowID 中；rowID 大于 0 表示插入数据成功，getContext(). getContentResolver(). notifyChange(uri, null)向 URI 的 Resolver 对象发送修改完成的通知。第 52~61 行重写 update(Uri, ContentValues, String, String[])方法，更新联系人的数据。第 64~84 行重写 delete(Uri, String, String[])方法，删除联系人的数据。第 87~93 行重写 query(Uri, String[], String, String[], String)方法，查找联系人。

3. 内容提供者主界面的 Activity

/java/com/vt/c0703/MainActivity.java

```
56   //自定义ContentObserver类
57   class ContactContentObserver extends ContentObserver {
58       public ContactContentObserver(Handler handler) {
59           super(handler);
60       }
61       //ContentProvider.notifyChange()方法后回调该方法
62       @Override
63       public void onChange(boolean selfChange) {
64           super.onChange(selfChange);
65           listAllContacts();
66       }
67   }
68   //初始化
69   private void init(){
70       //显示所有联系人
71       listAllContacts();
72       //注册观察者
73       ContactContentObserver contactContentObserver = new ContactContentObserver(new Handler());
74       ContentResolver contentResolver = this.getContentResolver();
75       contentResolver.registerContentObserver(URI,true,contactContentObserver);
76   }
77   //显示所有联系人
78   private void listAllContacts() {
79       mContactModel.clear();
80       //查询所有的联系人,按照id降序排列
81       Cursor cursor = mSQLiteDataBase.rawQuery("SELECT * FROM " + TABLE_NAME + " order by id desc", null);
82       //遍历查询结果指针
83       while (cursor.moveToNext()) {
```

```
84              String id = cursor.getString(cursor.getColumnIndex("id"));
85              String name = cursor.getString(cursor.getColumnIndex("name"));
86              String phone = cursor.getString(cursor.getColumnIndex("phone"));
87              mContactModel.add(new ContactModel(id, name, phone));
88          }
89          cursor.close();
90          //更新 ListView
91          mContactListAdapter.notifyDataSetChanged();
92      }
```

第 57～67 行 ContactContentObserver 继承自 ContentObserver，观察 ContentProvider 提供的数据变化；如果提供的数据发生变化，则调用 listAllContacts()方法重新查询显示联系人。第 73～75 行通过 ContentResolver 类对象注册 ContactContentObserver 类对象，观察指定 URI 提供的数据变化。第 78～92 行的 listAllContacts()方法首先清空 mContactModel 对象，然后查询到联系人的数据指针，使用遍历的方式获取每个联系人的数据并存储到 mContactModel 对象中，最后通过 mContactListAdapter.notifyDataSetChanged() 将 mContactModel 对象中存储的联系人数据更新显示到 ListView 控件中。

4．打开内容接收者的基础工程

打开"基础工程"文件夹中的"C0704"工程，该工程已经包含 MainActivity、ContactListAdapter、ContactModel 及布局的文件。ContactListAdapter 类是显示联系人的 ListView 控件的适配器类，ContactModel 类是联系人的模型类。

5．内容接收者主界面的 Activity

```
/java/com/vt/c0704/MainActivity.java
48      //显示所有联系人
49      private void listAllContacts() {
50          mContactModel.clear();
51          ContentResolver resolver = this.getContentResolver();
52          Uri uri = Uri.parse("content://com.vt.contact/contacts");
53          //查询 contacts 表的所有数据
54          Cursor cursor = resolver.query(uri, null, null, null, "name desc");
55          //遍历查询结果指针
56          while (cursor.moveToNext()) {
57              String idQuery = cursor.getString(cursor.getColumnIndex("id"));
58              String nameQuery = cursor.getString(cursor.getColumnIndex("name"));
59              String phoneQuery = cursor.getString(cursor.getColumnIndex("phone"));
60              mContactModel.add(new ContactModel(idQuery, nameQuery, phoneQuery));
61          }
62          cursor.close();
63          //更新 ListView
64          mContactListAdapter.notifyDataSetChanged();
65      }
66      //单击事件
67      @Override
68      public void onClick(View v) {
69          String id = mIdEditText.getText().toString();
```

```java
70      String name = mNameEditText.getText().toString();
71      String phone = mPhoneEditText.getText().toString();
72      Uri uri;
73      String selection;
74      String[] selectionArgs;
75      ArrayList<String> selectionList;
76      ContentResolver resolver = this.getContentResolver();
77      switch (v.getId()) {
78          case R.id.btn_list://显示所有联系人
79              listAllContacts();
80              break;
81          case R.id.btn_insert://添加联系人
82              uri = Uri.parse("content://com.vt.contact/contacts");
83              //将要插入的数据
84              ContentValues values = new ContentValues();
85              values.put("name", name);
86              values.put("phone", phone);
87              //插入数据
88              Uri resUri = resolver.insert(uri, values);
89              if (resUri == uri) {
90                  Toast.makeText(this, "插入失败", Toast.LENGTH_SHORT).show();
91              } else {
92                  listAllContacts();
93                  Toast.makeText(this, "插入成功", Toast.LENGTH_SHORT).show();
94              }
95              break;
96          case R.id.btn_query://查询联系人
97              mContactModel.clear();
98              uri = Uri.parse("content://com.vt.contact/contacts");
99              //合成查询数据条件的变量值
100             selection = (!TextUtils.isEmpty(id) ? "id=?" : "") + (!TextUtils.isEmpty(id)
    && !TextUtils.isEmpty(name) ? " AND " : "") + (!TextUtils.isEmpty(name) ? "name=?" : "")
    + ((!TextUtils.isEmpty(id) || !TextUtils.isEmpty(name)) && !TextUtils.isEmpty(phone) ?
    " AND " : "") + (!TextUtils.isEmpty(phone) ? "phone=?" : "");
101             selectionList = new ArrayList<>();
102             selectionArgs = new String[selectionList.size()];
103             if (!id.equals("")) selectionList.add(id);
104             if (!name.equals("")) selectionList.add(name);
105             if (!phone.equals("")) selectionList.add(phone);
106             selectionList.toArray(selectionArgs);
107             //查询数据
108             Cursor cursor = resolver.query(uri, null, selection, selectionArgs, "name desc");
109             //遍历查询结果指针
110             while (cursor.moveToNext()) {
111                 String idQuery = cursor.getString(cursor.getColumnIndex("id"));
112                 String nameQuery = cursor.getString(cursor.getColumnIndex("name"));
113                 String phoneQuery = cursor.getString(cursor.getColumnIndex("phone"));
114                 mContactModel.add(new ContactModel(idQuery, nameQuery, phoneQuery));
115             }
116             cursor.close();
```

```
117            mContactListAdapter.notifyDataSetChanged();
118            break;
119        case R.id.btn_update://更新联系人
120            if (!id.equals("")) {
121                uri = Uri.parse("content://com.vt.contact/contacts/" + id);
122                //将要更新的数据
123                ContentValues values1 = new ContentValues();
124                values1.put("name", name);
125                values1.put("phone", phone);
126                //更新数据
127                int resUpdate = resolver.update(uri, values1, null, null);
128                if (resUpdate > 0) {
129                    listAllContacts();
130                    Toast.makeText(this, "更新成功", Toast.LENGTH_SHORT).show();
131                } else {
132                    Log.d("aa", "更新失败");
133                    Toast.makeText(this, "更新失败", Toast.LENGTH_SHORT).show();
134                }
135            }
136            break;
137        case R.id.btn_delete://删除联系人
138            int resDelete = 0;
139            if (!id.equals("")) {
140                //根据id删除联系人
141                uri = Uri.parse("content://com.vt.contact/contacts/" + id);
142                resDelete = resolver.delete(uri, null, null);
143            } else if (!name.equals("")) {
144                //根据姓名和电话号码删除联系人
145                uri = Uri.parse("content://com.vt.contact/contacts");
146                resDelete = resolver.delete(uri, "name=?" + (!TextUtils.isEmpty(phone) ?
    " AND phone=?" : ""), new String[]{name, phone});
147            } else if (!phone.equals("")) {
148                //根据电话号码删除联系人
149                uri = Uri.parse("content://com.vt.contact/contacts");
150                resDelete = resolver.delete(uri, "phone=?", new String[]{phone});
151            }
152            if (resDelete > 0) {
153                listAllContacts();
154                Toast.makeText(this, "删除成功", Toast.LENGTH_SHORT).show();
155            } else {
156                Toast.makeText(this, "删除失败", Toast.LENGTH_SHORT).show();
157            }
158            break;
159        }
160    }
```

第 54 行通过 resolver.query(uri, null, null, null, "name desc") 获取联系人数据的指针。第 56~61 行遍历指针存储的联系人数据并显示出来。第 82~94 行通过 URI 向 C0703 发送添加联系人的数据请求，并处理返回结果。第 97~117 行清空 mContactModel 对象存储的联系人数据后，通过 URI 向 C0703 发送查询联系人的数据请求，并将查询到的联系人数据显示出来。第 120～135 行根据联系人的 id 通过 URI 向 C0703 发送更新联系人的数据请求，如果更新成功，则更新显示的联系人数据。第 138～157 行根据联系人的 id 通过 URI 向 C0703 发送删除联系人的数据请求，如果删除成功，则更新显示的联系人数据。

7.3.3 实例工程：访问和修改系统通讯录数据

本实例演示了使用 ContentResolver 访问和修改系统通讯录的联系人信息，首次运行时会打开获取访问系统通讯录权限的对话框。添加联系人后，能在系统通讯录中查看到添加的联系人信息（如图 7-8 所示）。

图 7-8 运行效果

> **提示：系统通讯录**
> 系统通讯录是系统自带的 App，使用 raw_contacts 表和 data 表保存联系人的数据。
> - raw_contacts 表：存储联系人。id 字段为主键，声明为 autoincrement，存储联系人的 id；display_name 字段存储联系人的姓名。对应的 URI 是 content://com.android.contacts/raw_contacts。
> - data 表：存储联系人具体的数据。raw_contact_id 字段存储联系人的 id，与 raw_contacts 表中联系人的 id 相对应；mimetype 字段存储记录数据的类型；data1～data9 字段存储具体的数据。对应的 URI 是 content://com.android.contacts/data。
>
> 每个联系人在 raw_contacts 表中只有一条记录；在 data 表中会有多条记录，每条记录保存一种形式的数据，通过 mimetype 字段进行区分。常用的 mimetype 字段值如下。
> - vnd.android.cursor.item/phone_v2：电话。
> - vnd.android.cursor.item/name：姓名。
> - vnd.android.cursor.item/email_v2：邮件。
> - vnd.android.cursor.item/postal-address_v2：地址。

- vnd.android.cursor.item/organization：组织。
- vnd.android.cursor.item/photo：照片。

1．打开基础工程

打开"基础工程"文件夹中的"C0705"工程，该工程已经包含 MainActivity、ContactListAdapter、ContactModel 及布局的文件。ContactListAdapter 类是显示联系人的 ListView 控件的适配器类，ContactModel 类是联系人的模型类。

提示：ContentProviderOperations 类

ContentProvider 类封装了数据的访问接口，其底层数据一般都保存在本地或服务器端。需要操作多行数据时，可以选择多次调用 ContentResolver 类的相关方法。为了使批量更新、插入、删除数据更加方便，Android 提供了 ContentProviderOperation 类，其优点如下：

- 在一个事务中执行所有操作，只需打开和关闭一个事务，能够保证数据完整性；
- 减少占用 CPU 的时间，提升性能，同时减少电量的消耗。

2．设置权限

```
/manifests/AndroidManifest.xml
02  <manifest xmlns:android="http://schemas.android.com/apk/res/android" package="com.vt.c0705">
03      <uses-permission android:name="android.permission.READ_CONTACTS"/>
04      <uses-permission android:name="android.permission.WRITE_CONTACTS"/>
```

第 03 行读取系统通讯录联系人的权限。第 04 行添加、删除和更新系统通讯录联系人的权限。

提示：动态权限

为了防止 Android 权限滥用危害用户信息安全，从 Android 6.0（API level 23）开始，在使用权限时不但需要在 AndroidManifest.xml 文件中设置权限，还要在使用权限时进行动态授权。动态权限牺牲了使用的便捷性，提升了系统的安全性。首次动态授权后，需在应用设置中对授权进行修改。目前，很多 App 强制用户开启权限，否则无法使用部分功能甚至全部功能。动态授权主要涉及以下三个方法。

- int ActivityCompat.checkSelfPermission（String permission）：检测是否已经获得该权限。
- void ActivityCompat.requestPermissions（String[] permissions, int requestCode）：请求单个或多个权限，此时会打开系统动态授权的对话框。
- void onRequestPermissionsResult（int requestCode, String[] permissions, int[] grantResults）：Activity 对请求权限结果的回调方法。

3．主界面的 Activity

```
/java/com/vt/c0705/MainActivity.java
25  public class MainActivity extends AppCompatActivity implements View.OnClickListener {
26      private static final int OPERATE_LIST = 0;
27      private static final int OPERATE_QUERY = 1;
28      private static final int OPERATE_INSERT = 2;
```

```java
29      private static final int OPERATE_UPDATE = 3;
30      private static final int OPERATE_DELETE = 4;
31      private static final Uri rawContactsUri = Uri.parse("content://com.android.contacts/raw_contacts");
32      private static final Uri dataUri = Uri.parse("content://com.android.contacts/data");
33      private List<ContactModel> mContactModel;
34      private ContactListAdapter mContactListAdapter;
35      private ListView mContactListView;
36      public EditText mIdEditText;
37      public EditText mNameEditText;
38      public EditText mPhoneEditText;
39      @Override
40      protected void onCreate(Bundle savedInstanceState) {
41          super.onCreate(savedInstanceState);
42          setContentView(R.layout.activity_main);
43          this.setTitle("C0705:访问和修改系统通讯录数据");
44          //初始化
45          mContactModel = new ArrayList<>();
46          mContactListAdapter = new ContactListAdapter(this, mContactModel);
47          mContactListView = findViewById(R.id.list_view_contacts);
48          mContactListView.setAdapter(mContactListAdapter);
49          mIdEditText = findViewById(R.id.edit_text_id);
50          mNameEditText = findViewById(R.id.edit_text_name);
51          mPhoneEditText = findViewById(R.id.edit_text_phone);
52          //设置监听器
53          findViewById(R.id.btn_list).setOnClickListener(this);
54          findViewById(R.id.btn_insert).setOnClickListener(this);
55          findViewById(R.id.btn_query).setOnClickListener(this);
56          findViewById(R.id.btn_update).setOnClickListener(this);
57          findViewById(R.id.btn_delete).setOnClickListener(this);
58          //判断是否获取读取联系人权限
59          if (ContextCompat.checkSelfPermission(this, Manifest.permission.READ_CONTACTS) != PackageManager.PERMISSION_GRANTED) {
60              //请求获取读取联系人权限
61              ActivityCompat.requestPermissions(this, new String[]{Manifest.permission.READ_CONTACTS}, OPERATE_LIST);
62          } else {
63              //显示所有联系人
64              listContacts();
65          }
66      }
67      //单击事件
68      @Override
69      public void onClick(View v) {
70          switch (v.getId()) {
71              case R.id.btn_list://显示所有联系人
72                  listContacts();
73                  break;
74              case R.id.btn_query://查询联系人
75                  //判断是否获取读取联系人权限
```

```java
76              if (ContextCompat.checkSelfPermission(this, Manifest.permission.READ_
   CONTACTS) != PackageManager.PERMISSION_GRANTED) {
77                  //请求获取读取联系人权限
78                  ActivityCompat.requestPermissions(this, new String[]{Manifest.permission.
   READ_CONTACTS}, OPERATE_QUERY);
79              } else {
80                  queryContact();
81              }
82              break;
83          case R.id.btn_insert://添加联系人
84              //判断是否获取写入联系人权限
85              if (ContextCompat.checkSelfPermission(this, Manifest.permission.WRITE_
   CONTACTS) != PackageManager.PERMISSION_GRANTED) {
86                  //请求获取写入联系人权限
87                  ActivityCompat.requestPermissions(this, new String[]{Manifest.
   permission.WRITE_CONTACTS}, OPERATE_INSERT);
88              } else {
89                  insertContact(true);
90              }
91              break;
92          case R.id.btn_update://更新联系人
93              //判断是否获取写入联系人权限
94              if (ContextCompat.checkSelfPermission(this, Manifest.permission.WRITE_
   CONTACTS) != PackageManager.PERMISSION_GRANTED) {
95                  //请求获取写入联系人权限
96                  ActivityCompat.requestPermissions(this, new String[]{Manifest.
   permission.WRITE_CONTACTS}, OPERATE_UPDATE);
97              } else {
98                  updateContact(true);
99              }
100             break;
101         case R.id.btn_delete://删除联系人
102             //判断是否获取写入联系人权限
103             if (ContextCompat.checkSelfPermission(this, Manifest.permission.WRITE_
   CONTACTS) != PackageManager.PERMISSION_GRANTED) {
104                 //请求获取写入联系人权限
105                 ActivityCompat.requestPermissions(this, new String[]{Manifest.permission.
   WRITE_CONTACTS}, OPERATE_DELETE);
106             } else {
107                 deleteContact(true);
108             }
109             break;
110         }
111     }
112     //权限请求的回调
113     @Override
114     public void onRequestPermissionsResult(int requestCode, String[] permissions,
   int[] grantResults) {
115         switch (requestCode) {
116             case OPERATE_LIST:
```

```java
117             if (grantResults.length > 0 && grantResults[0] == PackageManager.PERMISSION_GRANTED) {
118                 listContacts();
119             } else {
120                 Toast.makeText(this, "您没有读取联系人的权限", Toast.LENGTH_SHORT).show();
121             }
122             break;
123         case OPERATE_QUERY:
124             if (grantResults.length > 0 && grantResults[0] == PackageManager.PERMISSION_GRANTED) {
125                 queryContact();
126             } else {
127                 Toast.makeText(this, "您没有查询联系人的权限", Toast.LENGTH_SHORT).show();
128             }
129             break;
130         case OPERATE_INSERT:
131             if (grantResults.length > 0 && grantResults[0] == PackageManager.PERMISSION_GRANTED) {
132                 insertContact(true);
133             } else {
134                 Toast.makeText(this, "您没有添加联系人的权限", Toast.LENGTH_SHORT).show();
135             }
136             break;
137         case OPERATE_UPDATE:
138             if (grantResults.length > 0 && grantResults[0] == PackageManager.PERMISSION_GRANTED) {
139                 updateContact(true);
140             } else {
141                 Toast.makeText(this, "您没有修改联系人的权限", Toast.LENGTH_SHORT).show();
142             }
143             break;
144         case OPERATE_DELETE:
145             if (grantResults.length > 0 && grantResults[0] == PackageManager.PERMISSION_GRANTED) {
146                 deleteContact(true);
147             } else {
148                 Toast.makeText(this, "您没有删除联系人的权限", Toast.LENGTH_SHORT).show();
149             }
150             break;
151         default:
152         }
153     }
154     //查询所有联系人
155     private void listContacts() {
156         mContactModel.clear();
157         ContentResolver resolver = getContentResolver();
158         //查询raw_contacts表中存储的联系人
159         Cursor rawContactsCursor = resolver.query(rawContactsUri, null, null, null, null);
160         //遍历查询结果
161         while (rawContactsCursor.moveToNext()) {
162             //获取raw_contacts表中存储的联系人id和显示名字
```

```
163            String rawContactId = rawContactsCursor.getString(
    rawContactsCursor.getColumnIndex(ContactsContract.CommonDataKinds.Phone._ID));
164            String displayname = rawContactsCursor.getString(
    rawContactsCursor.getColumnIndex(ContactsContract.CommonDataKinds.Phone.DISPLAY_NAME));
165            //根据联系人id查询data表中存储的联系人电话号码
166            Cursor dataCursor = resolver.query(dataUri, null, "raw_contact_id=? AND
    mimetype='vnd.android.cursor.item/phone_v2'", new String[]{rawContactId}, null);
167            while (dataCursor.moveToNext()) {
168                String phone = dataCursor.getString(dataCursor.getColumnIndex("data1"));
169                mContactModel.add(new ContactModel(rawContactId, displayname, phone));
170            }
171        }
172        rawContactsCursor.close();
173        mContactListAdapter.notifyDataSetChanged();
174    }
175    //根据显示名字查询联系人
176    private void queryContact() {
177        String name = mNameEditText.getText().toString().trim();
178        ContentResolver resolver = getContentResolver();
179        //根据联系人的显示名字搜索
180        Cursor rawContactsCursor = resolver.query(rawContactsUri, null, "display_name=?",
    new String[]{name}, null);
181        //判断是否搜索到联系人
182        if (rawContactsCursor.getCount() > 0) {
183            mContactModel.clear();
184            while (rawContactsCursor.moveToNext()) {
185                //获取raw_contacts表中存储的联系人id和显示名字
186                String rawContactId = rawContactsCursor.getString(
    rawContactsCursor.getColumnIndex(ContactsContract.CommonDataKinds.Phone._ID));
187                String displayname = rawContactsCursor.getString(
    rawContactsCursor.getColumnIndex(ContactsContract.CommonDataKinds.Phone.DISPLAY_NAME));
188                //根据联系人id查询data表中存储的联系人电话号码
189                Cursor dataCursor = resolver.query(dataUri, null, "raw_contact_id=? AND
    mimetype='vnd.android.cursor.item/phone_v2'", new String[]{rawContactId}, null);
190                while (dataCursor.moveToNext()) {
191                    String phone = dataCursor.getString(dataCursor.getColumnIndex("data1"));
192                    mContactModel.add(new ContactModel(rawContactId, displayname, phone));
193                }
194            }
195            Toast.makeText(this, "已经查询到该联系人", Toast.LENGTH_SHORT).show();
196            mContactListAdapter.notifyDataSetChanged();
197        } else {
198            Toast.makeText(this, "没有查询到该联系人", Toast.LENGTH_SHORT).show();
199        }
200        rawContactsCursor.close();
201    }
202    //添加联系人
203    private void insertContact(boolean useContentProviderOperation) {
204        String name = mNameEditText.getText().toString();
205        String phone = mPhoneEditText.getText().toString();
```

```java
206         //是否使用ContentProviderOperation
207         if (useContentProviderOperation) {
208             //向raw_contacts表中插入联系人的操作
209             ArrayList<ContentProviderOperation> operations = new ArrayList<>();
210             ContentProviderOperation op1 = ContentProviderOperation.newInsert(rawContactsUri)
211                     .withValue("account_name", null)
212                     .build();
213             operations.add(op1);
214             //向data表中插入联系人的显示名字的操作
215             ContentProviderOperation op2 = ContentProviderOperation.newInsert(dataUri)
216                     .withValueBackReference("raw_contact_id", 0)
217                     .withValue("mimetype", "vnd.android.cursor.item/name")
218                     .withValue("data2", name)
219                     .build();
220             operations.add(op2);
221             //向data表中插入联系人的电话号码的操作
222             ContentProviderOperation op3 = ContentProviderOperation.newInsert(dataUri)
223                     .withValueBackReference("raw_contact_id", 0)
224                     .withValue("mimetype", "vnd.android.cursor.item/phone_v2")
225                     .withValue("data1", phone)
226                     .build();
227             operations.add(op3);
228             try {
229                 ContentResolver resolver = getContentResolver();
230                 ContentProviderResult[] contentProviderResult = resolver.applyBatch("com.android.contacts", operations);
231                 //判断是否添加成功
232                 if (contentProviderResult.length == operations.size()) {
233                     Toast.makeText(this, "添加联系人成功", Toast.LENGTH_SHORT).show();
234                     listContacts();
235                 } else {
236                     Toast.makeText(this, "添加联系人失败", Toast.LENGTH_SHORT).show();
237                 }
238             } catch (OperationApplicationException e) {
239                 e.printStackTrace();
240                 Toast.makeText(this, "添加联系人异常: " + e, Toast.LENGTH_SHORT).show();
241             } catch (RemoteException e) {
242                 e.printStackTrace();
243                 Toast.makeText(this, "添加联系人异常: " + e, Toast.LENGTH_SHORT).show();
244             }
245         } else {
246             ContentResolver resolver = getContentResolver();
247             //向raw_contacts表中插入联系人的操作
248             ContentValues value1 = new ContentValues();
249             value1.putNull("account_name");
250             Uri resultUri = resolver.insert(rawContactsUri, value1);
251             int rawContactsId = Integer.valueOf(resultUri.getPathSegments().get(1));
252             //向data表中插入联系人的显示名字的操作
253             ContentValues value2 = new ContentValues();
254             value2.put("raw_contact_id", rawContactsId);
```

```
255            value2.put("mimetype", "vnd.android.cursor.item/name");
256            value2.put("data2", name);
257            Uri dataNameUri = resolver.insert(dataUri, value2);
258            int dataNameId = Integer.valueOf(dataNameUri.getPathSegments().get(1));
259            //向data表中插入联系人的电话号码的操作
260            ContentValues value3 = new ContentValues();
261            value3.put("raw_contact_id", rawContactsId);
262            value3.put("mimetype", "vnd.android.cursor.item/phone_v2");
263            value3.put("data1", phone);
264            Uri dataPhoneUri = resolver.insert(dataUri, value3);
265            int dataPhoneId = Integer.valueOf(dataPhoneUri.getPathSegments().get(1));
266            //判断是否添加成功
267            if (rawContactsId > 0 && dataNameId > 0 && dataPhoneId > 0) {
268                Toast.makeText(this, "添加联系人成功", Toast.LENGTH_SHORT).show();
269                listContacts();
270            } else {
271                Toast.makeText(this, "添加联系人失败或部分失败", Toast.LENGTH_SHORT).show();
272            }
273        }
274    }
275    //修改联系人
276    private void updateContact(boolean useContentProviderOperation) {
277        String id = mIdEditText.getText().toString().trim();
278        String name = mNameEditText.getText().toString().trim();
279        String phone = mPhoneEditText.getText().toString().trim();
280        //是否使用ContentProviderOperation
281        if(useContentProviderOperation){
282            ArrayList<ContentProviderOperation> operations = new ArrayList<>();
283            //根据联系人的id更新raw_contacts表的操作
284            ContentProviderOperation op1 = ContentProviderOperation.newUpdate(rawContactsUri)
285                    .withValue("display_name", name)
286                    .withSelection("_id=?", new String[]{id})
287                    .build();
288            operations.add(op1);
289            //根据联系人的id更新data表
290            ContentProviderOperation op2 = ContentProviderOperation.newUpdate(dataUri)
291                    .withValue("data1", phone)
292                    .withSelection("raw_contact_id=? AND mimetype='vnd.android.cursor.item/phone_v2'", new String[]{id})
293                    .build();
294            operations.add(op2);
295            try {
296                ContentResolver resolver = getContentResolver();
297                //批量执行操作，并将结果返回
298                ContentProviderResult[] contentProviderResult = resolver.applyBatch("com.android.contacts", operations);
299                //判断是否添加成功
300                if (contentProviderResult.length == operations.size()) {
301                    Toast.makeText(this, "修改联系人成功", Toast.LENGTH_SHORT).show();
302                    listContacts();
```

```java
303            } else {
304                Toast.makeText(this, "修改联系人失败", Toast.LENGTH_SHORT).show();
305            }
306        } catch (OperationApplicationException e) {
307            e.printStackTrace();
308            Toast.makeText(this, "修改联系人异常: " + e, Toast.LENGTH_SHORT).show();
309        } catch (RemoteException e) {
310            e.printStackTrace();
311            Toast.makeText(this, "修改联系人异常: " + e, Toast.LENGTH_SHORT).show();
312        }
313    }else{
314        ContentResolver resolver = getContentResolver();
315        //raw_contacts 表将要更新的数据
316        ContentValues values1 = new ContentValues();
317        values1.put("display_name", name);
318        //根据联系人的 id 更新 raw_contacts 表
319        int resUpdateName = resolver.update(rawContactsUri, values1, "_id=?", new String[]{id});
320        //data 表将要更新的数据
321        ContentValues values2 = new ContentValues();
322        values2.put("data1", phone);
323        //根据联系人的 id 更新 data 表
324        int resUpdatePhone = resolver.update(dataUri, values2, "raw_contact_id=? AND mimetype='vnd.android.cursor.item/phone_v2'", new String[]{id});
325        if (resUpdateName > 0 && resUpdatePhone > 0) {
326            Toast.makeText(this, "修改联系人成功", Toast.LENGTH_SHORT).show();
327            listContacts();
328        } else {
329            Toast.makeText(this, "修改联系人失败或部分失败", Toast.LENGTH_SHORT).show();
330        }
331    }
332  }
333  //删除联系人
334  private void deleteContact(boolean useContentProviderOperation) {
335      String idDelete = mIdEditText.getText().toString();
336      //是否使用 ContentProviderOperation
337      if(useContentProviderOperation){
338          ArrayList<ContentProviderOperation> operations = new ArrayList<>();
339          //删除 raw_contacts 表中的联系人数据的操作
340          ContentProviderOperation op1 = ContentProviderOperation.newDelete(rawContactsUri)
341                  .withSelection("_id=?", new String[]{idDelete})
342                  .build();
343          operations.add(op1);
344          //删除 data 表中的联系人数据的操作
345          ContentProviderOperation op2 = ContentProviderOperation.newDelete(dataUri)
346                  .withSelection("raw_contact_id=?", new String[]{idDelete})
347                  .build();
348          operations.add(op2);
349          try {
350              ContentResolver resolver = getContentResolver();
351              //批量执行操作，并将结果返回
```

```
352                ContentProviderResult[] contentProviderResult = resolver.applyBatch("com.
       android.contacts", operations);
353                //判断是否添加成功
354                if (contentProviderResult.length == operations.size()) {
355                    Toast.makeText(this, "删除联系人成功", Toast.LENGTH_SHORT).show();
356                    listContacts();
357                } else {
358                    Toast.makeText(this, "删除联系人失败", Toast.LENGTH_SHORT).show();
359                }
360            } catch (OperationApplicationException e) {
361                e.printStackTrace();
362                Toast.makeText(this, "删除联系人异常: " + e, Toast.LENGTH_SHORT).show();
363            } catch (RemoteException e) {
364                e.printStackTrace();
365                Toast.makeText(this, "删除联系人异常: " + e, Toast.LENGTH_SHORT).show();
366            }
367        }else{
368            ContentResolver resolver = getContentResolver();
369            //删除 raw_contacts 表中的联系人数据，返回值大于 0 说明有被删除的数据
370            int rawContactsDeleteResult = resolver.delete(rawContactsUri, "_id=?", new
       String[]{idDelete});
371            //删除 data 表中的联系人数据，返回值大于 0 说明有被删除的数据
372            int dataDeleteResult = resolver.delete(dataUri, "raw_contact_id=?", new
       String[]{idDelete});
373            //判断是否删除成功
374            if(rawContactsDeleteResult>0 && dataDeleteResult>0){
375                Toast.makeText(this, "删除联系人成功", Toast.LENGTH_SHORT).show();
376                listContacts();
377            }else{
378                Toast.makeText(this, "删除联系人失败或部分失败", Toast.LENGTH_SHORT).show();
379            }
380        }
381    }
382 }
```

第 59～65 行请求读取系统联系人的权限，如果未获取该权限，则打开请求权限的对话框；如果已经获取该权限，则调用 listContacts()方法。第 69～111 行单击按钮时需要先判断是否已经获取相应的权限，以避免无权限所产生的错误。第 114～153 行请求权限的对话框关闭后的回调方法，判断用户的授权结果，如果取得授权，则继续执行调用相应的方法。第 155～174 行查询系统通讯录中所有的联系人并显示出来。第 176～201 行根据联系人的名字查询系统通讯录中是否存在此人，如果存在则显示查询结果。第 207～244 行使用 ContentProviderOperation 的方式向系统通讯录中添加联系人，这种方式更便于理解系统联系人的数据表结构。第 247～274 行使用 ContentValues 的方式向系统通讯录中添加联系人，这种方式比较简捷。第 275～332 行使用 ContentProviderOperation 和 ContentValues 方式修改系统通讯录中的联系人。第 333～381 行使用 ContentProviderOperation 和 ContentValues 方式删除系统通讯录中的联系人。

7.4 JavaScript 对象表示法

7.4.1 JSON 概述

JSON（JavaScript Object Notation）是一种轻量级的数据交换方法，是基于 JavaScript 的一个子集，主要用于网络或程序之间传递数据。JSON 使用字符串形式保存数据，结构层次简捷清晰，易于解析和生成，占用空间小，有效地提升网络传输效率。通常情况下，JSON 数据并不会直接呈现给用户，而是在后台与服务器端进行数据通信。

JSON 对象是 JSON 的基本构成单位，JSON 数组用于存储一类 JSON 对象集合（如表 7-12 所示）。JSON 对象表示为键值对，数据由逗号分隔，花括号保存对象，使用双引号保存键名和键值，键名和键值之间使用冒号分隔。如果值是 String 类型且含有双引号或冒号，则需要使用"\"转义。

表 7-12 JSON 对象和 JSON 数组的实例对比

JSON 对象	JSON 数组
{"name": "小白魔", "content": "今天你吃了吗？"}	[{"name": "小白魔", "content": "今天你吃了吗？"}, {"name": "天使", "content": "关你什么事！"}]

XML（Extensible Markup Language）是可扩展标记语言，是标准通用标记语言的子集。XML 诞生于 1998 年，早于 2001 年诞生的 JSON。JSON 与 XML 相比（如表 7-13 所示），数据的描述性较差，数据可读性基本相同，数据的体积更小，传输与解析速度更快。

表 7-13 XML 与 JSON 数据实例对比

XML 数据	JSON 数据

多行形式：
<?xml version="1.0" encoding="utf-8"?>
<country>
  <name>中国</name>
  <province>
    <name>辽宁</name>
    <cities>
      <city>沈阳</city>
      <city>本溪</city>
    </cities>
  </province>
  <province>
    <name>新疆</name>
    <cities>
      <city>乌鲁木齐</city>
      <city>喀什</city>
    </cities>
  </province>
</country>
	多行形式：
	{
	  "name": "中国",
	  "province": [
	    {
	      "name": "辽宁",
	      "cities": {"city": ["沈阳", "本溪"]}
	    },
	    {
	      "name": "新疆",
	      "cities": {"city": ["乌鲁木齐", "喀什"]}
	    }
	  ]
	}

续表

XML 数据	JSON 数据
单行形式： <?xml version="1.0" encoding="utf-8"?><country><name> 中国 </name><province><name> 辽宁 </name><cities><city> 沈阳 </city><city> 本溪 </city></cities></province><province><name> 新疆 </name><cities><city> 乌鲁木齐 </city><city> 喀什 </city></cities></province></country>	单行形式： {"name": "中国","province": [{"name": "辽宁","cities": {"city": ["沈阳", "本溪"]}}, {"name": "新疆","cities": {"city": ["乌鲁木齐", "喀什"]}}]}

JSON 有很多支持库，常见的包括 GSON、FastJSON 和 Jackson。Android 自带的 JSON 库是 org.json，可以直接使用，能够满足大多数的使用需求，比第三方库更加方便，提供了 JSONObject、JSONArray、JSONStringer 和 JSONTokener 四个类。实际使用时推荐直接使用 JSONObject、JSONArray 合成和解析 JSON，解析后赋值给 Array 或 ArrayList 对象再进行其他操作。

JSONObject 类（org.json.JSONObject）用于新建、读取、解析和操作 JSON 对象（如表 7-14 所示）。键名是非空字符串，键值可以是 JSONObject、JSONArray、String、boolean、int、long、double 或 NULL 的任意组合。调用 putXXX(name,null) 将从对象中移除键名，但是调用 putXXX(name, JSONObject.null) 会存储 JSONObject.null 值，getXXX() 方法失败时返回 JSONObject.null。JSONArray（org.json.JSONArray）与 JSONObject 所提供的方法类似，但没有提供 names() 方法，putXXX() 方法的返回值类型为 JSONArray。

表 7-14　JSONObject 类的常用方法

类型和修饰符	方　　法
	JSONObject() 构造方法
	JSONObject(String json) 构造方法，通过 JSON 字符串实例化
Object	get(String name) 根据键名获取 Object 型键值，如果不存在则抛出异常
boolean	getBoolean(String name) 根据键名获取 boolean 型键值，如果不存在则抛出异常
double	getDouble(String name) 根据键名获取 double 型键值，如果不存在则抛出异常
int	getInt(String name) 根据键名获取 int 型键值，如果不存在则抛出异常
JSONArray	getJSONArray(String name) 根据键名获取 JSONArray 型键值，如果不存在则抛出异常
JSONObject	getJSONObject(String name) 根据键名获取 JSONObject 型键值，如果不存在则抛出异常
long	getLong(String name) 根据键名获取 long 型键值，如果不存在则抛出异常
String	getString(String name) 根据键名获取 String 型键值，如果不存在则抛出异常
boolean	has(String name) 判断是否有包含该键名的数据

续表

类型和修饰符	方法
boolean	isNull(String name) 判断键值是否为 NULL
int	length() 返回包含对象数量
JSONArray	names() 返回包含键名的 JSONArray 对象
Object	opt(String name) 根据键名获取 String 型键值，如果不存在则返回 null
boolean	optBoolean(String name, boolean fallback) 根据键名获取 boolean 型或转换为 boolean 型后的键值，否则返回 fallback
double	optDouble(String name, double fallback) 根据键名获取 double 型或转换为 double 型后的键值，否则返回 fallback
int	optInt(String name, int fallback) 根据键名获取 int 型或转换为 int 型后的键值，否则返回 fallback
JSONArray	optJSONArray(String name) 根据键名获取 JSONArray 型键值，否则返回 null
JSONObject	optJSONObject(String name) 根据键名获取 JSONObject 型键值，否则返回 null
long	optLong(String name, long fallback) 根据键名获取 long 型或转换为 long 型后的键值，否则返回 fallback
String	optString(String name, String fallback) 根据键名获取 String 型或转换为 String 型后的键值，否则返回 fallback
JSONObject	put(String name, double value) 将 double 型键值存入 name 键名中，能够覆盖该键名之前存储的键值
JSONObject	put(String name, boolean value) 将 boolean 型键值存入 name 键名中，能够覆盖该键名之前存储的键值
JSONObject	put(String name, int value) 将 int 型键值存入 name 键名中，能够覆盖该键名之前存储的键值
JSONObject	put(String name, long value) 将 long 型键值存入 name 键名中，能够覆盖该键名之前存储的键值
JSONObject	put(String name, Object value) 将 Object 型键值存入 name 键名中，能够覆盖该键名之前存储的键值
JSONObject	putOpt(String name, Object value) 当两个参数都非空时，与 put(name, value)等价；否则不执行任何操作
Object	remove(String name) 移除键值，否则不执行任何操作
JSONArray	toJSONArray(JSONArray names) 返回值与名称对应的数组
String	toString() 转换为单行的 JSON 字符串
String	toString (int indentSpaces) 转换为多行的 JSON 字符串，indentSpaces 表示每层嵌套缩进的空格数

7.4.2 实例工程：合成和解析 JSON 数据

本实例模拟了发布朋友圈或微博动态前合成 JSON（如图 7-9 所示），实例中没有演示合成后数据发往服务器端的过程。从服务器端获取到朋友圈或微博动态的 JSON 数据后，对JSON数据进行解析并显示，实例中没有演示从服务器端获取JSON数据的过程，而使用合成后的数据直接进行解析。单击"合成单行"或"合成多行"按钮，会在下方显示合成后的单行或多行JSON字符串。单击"清空"按钮，清空顶部EditText控件的内容。单击"解析"按钮，将合成的JSON数据解析后显示在顶部相应的EditText控件中。

图 7-9　运行效果

1．打开基础工程

打开"基础工程"文件夹中的"C0706"工程，该工程已经包含 MainActivity 及布局的文件。

2．主界面的 Activity

/java/com/vt/c0706/MainActivity.java	
47	`　　//合成 JSON`
48	`　　private void encodeJSON(boolean isSingle) {`
49	`　　　　JSONObject jsonObject = new JSONObject();`
50	`　　　　try {`
51	`　　　　　　//将昵称和内容存入 jsonObject 对象`
52	`　　　　　　jsonObject.put("name", mNameEditText.getText());`
53	`　　　　　　jsonObject.put("content", mContentEditText.getText());`
54	`　　　　　　//将图像分别存入新建的 JSONObject 类对象中`
55	`　　　　　　JSONObject pic0JsonObject = new JSONObject();`
56	`　　　　　　pic0JsonObject.put("type", "pic");`
57	`　　　　　　pic0JsonObject.put("url", mExtraEditText[0].getText());`
58	`　　　　　　JSONObject pic1JsonObject = new JSONObject();`
59	`　　　　　　pic1JsonObject.put("type", "pic");`
60	`　　　　　　pic1JsonObject.put("url", mExtraEditText[1].getText());`
61	`　　　　　　JSONObject pic2JsonObject = new JSONObject();`
62	`　　　　　　pic2JsonObject.put("type", "pic");`

```
63              pic2JsonObject.put("url", mExtraEditText[2].getText());
64              //将存入图像的三个 JSONObject 类对象存入 jsonArray 对象
65              JSONArray jsonArray = new JSONArray();
66              jsonArray.put(pic0JsonObject);
67              jsonArray.put(pic1JsonObject);
68              jsonArray.put(pic2JsonObject);
69              //将保存图像存入 jsonObject 对象
70              jsonObject.put("extra", jsonArray);
71              //判断是单行还是多行显示
72              if (isSingle) {
73                  mResultTextView.setText(jsonObject.toString());
74              } else {
75                  mResultTextView.setText(jsonObject.toString(4));
76              }
77          } catch (JSONException e) {
78              e.printStackTrace();
79          }
80      }
81      //解析 JSON
82      private void decodeJSON() {
83          try {
84              JSONObject jsonObject = new JSONObject(mResultTextView.getText().toString().trim());
85              //获取 name 键值的数据
86              String name = jsonObject.getString("name");
87              mNameEditText.setText(name);
88              //获取 content 键值的数据
89              String content = jsonObject.getString("content");
90              mContentEditText.setText(content);
91              //获取 extra 键值的数据
92              JSONArray jsonArray = jsonObject.getJSONArray("extra");
93              //遍历 extra 存储的 JSONArray 数据
94              for (int i = 0; i < jsonArray.length(); i++) {
95                  JSONObject extraJSONObject = jsonArray.getJSONObject(i);
96                  mExtraEditText[i].setText(extraJSONObject.optString("url", "images/default.jpg"));
97              }
98              //额外知识：Logcat 中递归自动解析 JSON 字符串的所有数据
99              iterateJSON(jsonObject, "");
100         } catch (JSONException e) {
101             e.printStackTrace();
102         }
103     }
```

第 52、53 行将 name 和 content 元素存储在 jsonObject 对象中。第 55~63 行新建 pic0JsonObject、pic1JsonObject 和 pic2JsonObject 对象，分别存储三组 type 和 url 元素。第 65~68 行新建 jsonArray 对象，然后将 pic0JsonObject、pic1JsonObject 和 pic2JsonObject 存入。第 70 行将 jsonArray 对象存入 jsonObject 对象的 extra 元素中。第 86 行 jsonObject.getString("name")解析出 String 类型的 name 元素。第 92~97 行通过 jsonObject.getJSONArray("extra") 解析出 JSONArray 类型的 extra 元素，并遍历 extra 元素中的所有子元素。

7.5 习　　题

1．使用 SharedPreferences 实现自动登录的功能。如果 24 小时内登录过，可以自动登录，否则重新登录。

2．使用 SQLite 实现记录登录时间的功能，并且能够查询登录的历史记录。

3．实现商品信息的 JSON 数据合成和解析，商品信息至少包含商品名称、价格、剩余数量。

第 8 章　多媒体与传感器

除了可以通过触摸屏和网络、外部进行交互，还可以使用摄像头、麦克风和传感器获取外部的信息。从本质上看，摄像头和麦克风也属于传感器，但是 Android 系统将两者独立出来，不仅是因为二者功能丰富，还因为它们是用户最能直接感知到的传感器。因此 Android 系统提供了独立的两个 App——相机和录音机。

8.1　系统相机和相册

Android 系统预置了相机和相册两个 App，可以通过 Intent 对象调用这两个 App，进行拍照、录制视频、选取图片和选取视频操作。微信、微博、小红书、抖音、大众点评等并没有使用内置的功能选择照片或视频，而是先通过 ContentProvider 获取相册中的照片数据并重新显示出来，再实现多选的功能，甚至还可以实现添加标签和编辑功能。

8.1.1　实例工程：拍照、选取和显示图片

本实例演示了使用系统相机 App 拍照及使用系统相册选取图片。首次运行时会打开权限请求的系统对话框，单击"拒绝"按钮打开自定义的对话框提示需要开启的权限，再单击"打开应用设置"按钮（如图 8-1 所示）。单击"拍照"按钮返回的拍摄照片画质较低，不适合实际使用；单击"拍照（指定位置保存）"按钮返回的拍摄照片画质正常，保存在指定文件夹中，并且能够显示在系统相册中；单击"选取图片"按钮选取照片，进行剪裁后显示方形图片（如图 8-2 所示）。调用系统相机拍照不会自动将照片保存在系统相册中，需要创建保存照片的文件，扫描媒体文件并发送广播才会显示在系统相册中。

图 8-1　权限请求和手动设置权限的运行效果

图 8-2 拍照和选取图片的运行效果

1．打开基础工程

打开"基础工程"文件夹中的"C0801"工程，该工程已经包含 MainActivity 及布局的文件。

2．Util 类

在"/java/com/vt/c0801"文件夹中，新建 Util 类，用于存储几种能够重复使用的公有方法。

/java/com/vt/c0801/Util.java

```java
12  public class Util {
13      //创建文件
14      public static File createFile(String parentPath, String childPath, String extension) {
15          //使用时间生成文件名
16          SimpleDateFormat format = new SimpleDateFormat("yyyyMMdd_hhmmss");
17          String fileName = format.format(new Date());
18          //存储路径
19          File storageDir = new File(parentPath, childPath);
20          if (!storageDir.exists()) {
21              storageDir.mkdirs();
22          }
23          //创建文件
24          File saveRecorderFile = new File(storageDir, "IMG_" + fileName + "." + extension);
25          if (saveRecorderFile.exists()) {
26              saveRecorderFile.delete();
27          }
28          return saveRecorderFile;
29      }
30      //将图片或视频显示在系统相册中
31      public static void showInAlbum(Context context, String path) {
32          //Android 中分割字符串需要在分隔符左右两侧加上中括号
33          String[] str = path.split("[.]");
```

```
34            //获取扩展名对应的文件类型值
35            String    mimeType    =    MimeTypeMap.getSingleton().getMimeTypeFromExtension
(str[str.length - 1]);
36            //根据路径和扩展名的类型扫描媒体文件
37            MediaScannerConnection.scanFile(context, new String[]{path}, new String[]{mimeType}, null);
38            //将图片显示在系统相册中
39            context.sendBroadcast(new Intent(Intent.ACTION_MEDIA_SCANNER_SCAN_FILE, Uri.parse
("file://" + path)));
40        }
41 }
```

第 14 行 createFile 方法创建存储照片的文件，设置了三个参数，parentPath 表示存储新建文件的父路径，childPath 表示存储新建文件的子路径，extension 表示新建文件的扩展名。第 19 行为了增加控制的灵活性，使用父路径和子路径合成存储照片的文件夹路径，即 parentPath+childPath。第 24 行创建保存图片的文件，文件名的前缀为"IMG_"。第 35 行 str[str.length - 1]获取文件的扩展名。

3．FileProvider 的权限路径

在"/res"文件夹中，新建"xml"文件夹。然后在"/res/xml"文件夹中，新建"file_paths.xml"文件，用于设置 FileProvider 的权限路径。

```
/res/xml/file_paths.xml
01  <?xml version="1.0" encoding="utf-8"?>
02  <paths>
03      <external-media-path name="cache_images" path="C0801"/>
04  </paths>
```

第 03 行表示路径为 Context.getExternalMediaDirs()+"C0801"，Context.getExternalMediaDirs() 的文件夹可以添加到系统相册中并进行显示。

 提示：FileProvider

Android 7.0（API level 24）禁止对外部（跨越应用分享）公开 file://，若使用 file:// 格式共享文件，会抛出 FileUriExposedException 异常，因此需要使用 FileProvider 类生成 content://类型的 URI 分享，并且为其提供临时的文件访问权限。

在 AndroidManifest.xml 文件中使用<provider>标签声明访问路径，具体的路径保存在 XML 文件中。在 XML 文件中使用<paths>作为顶层标签，<paths>子标签指定文件夹。<paths>子标签的 name 属性是 path 属性指定路径的替代名称，用于隐藏真实的目录；<paths>子标签的 path 属性表示指定路径下的共享目录。常用的<paths>子标签如下。

- <files-path>表示 Context.getFilesDir()所指向的目录。
- <cache-path>表示 Context.getCacheDir()所指向的目录。
- <external-path>表示 Environment.getExternalStorageDirectory()所指向的目录。
- <external-files-path>表示 Context.getExternalFilesDir()所指向的目录。
- <external-cache-path>表示 Context.getExternalCacheDir()所指向的目录。
- <external-media-path>表示 Context.getExternalMediaDirs()所指向的目录（从 API level 21 版本开始支持）。

4．Permissions 类

在"/java/com/vt/c0801"文件夹中，新建 Permissions 类，用于封装权限申请，便于重复使用。

```
/java/com/vt/c0801/Permissions.java
12  public class Permissions {
13      //权限请求码
14      public static final int REQUEST_PERMISSIONS = 1;
15      //是否已经获取权限
16      public static boolean hasPermissionsGranted(Context context, String[] permissions) {
17          for (String permission : permissions) {
18              //只要有一个拒绝的申请权限，就返回 false
19              if (ActivityCompat.checkSelfPermission(context, permission)
                          != PackageManager.PERMISSION_GRANTED) {
20                  return false;
21              }
22          }
23          return true;
24      }
25      //请求权限
26      public static void requestVideoPermissions(final Context context, String[] permissions) {
27          //如果已经拒绝了权限申请，打开对话框提示手动开启权限，否则显示权限请求的系统对话框
28          if (shouldShowRequestPermissionRationale(context, permissions)) {
29              String msg = "";
30              for (int i = 0; i < permissions.length; i++) {
31                  if (i == 0) {
32                      msg = permissions[i];
33                  } else {
34                      msg = msg + "\n" + permissions[i];
35                  }
36              }
37              //自定义对话框
38              AlertDialog.Builder adBuilder = new AlertDialog.Builder(context);
39              adBuilder.setIcon(R.mipmap.ic_launcher);
40              adBuilder.setTitle("需要手动开启以下权限");
41              adBuilder.setMessage(msg);
42              //单击确认按钮事件
43              adBuilder.setPositiveButton("打开应用设置", new DialogInterface.OnClickListener() {
44                  @Override
45                  public void onClick(DialogInterface dialog, int which) {
46                      //打开应用设置
47                      Intent intent = new Intent();
48                      intent.setAction("android.settings.AppLICATION_DETAILS_SETTINGS");
49                      intent.setData(Uri.fromParts("package", context.getPackageName(), null));
50                      context.startActivity(intent);
51                      dialog.dismiss();
52                  }
53              });
54              adBuilder.show();
```

```
55              } else {
56                  //显示权限请求的系统对话框
57                  ActivityCompat.requestPermissions((Activity) context, permissions, REQUEST_PERMISSIONS);
58              }
59          }
60          //获取是否拒绝过权限请求
61          public static boolean shouldShowRequestPermissionRationale(Context context,
    String[] permissions) {
62              for (String permission : permissions) {
63                  //如果有拒绝的申请权限，就返回 true
64                  if (ActivityCompat.shouldShowRequestPermissionRationale((Activity) context, permission)) {
65                      return true;
66                  }
67              }
68              return false;
69          }
70      }
```

第 16 行 permissions 参数传递申请的权限。第 19 行 PackageManager.PERMISSION_GRANTED 表示已经获取权限。第 26 行 final 修饰符可以使 context 用于第 49 行的监听事件中。

5．设置权限和 FileProvider 的权限路径

```
/manifests/AndroidManifest.xml
03   <uses-permission android:name="android.permission.CAMERA" />
04   <uses-permission android:name="android.permission.READ_EXTERNAL_STORAGE" />
05   <uses-permission android:name="android.permission.WRITE_EXTERNAL_STORAGE" />
19       <provider
20           android:name="androidx.core.content.FileProvider"
21           android:authorities="com.vt.c0801.fileprovider"
22           android:exported="false"
23           android:grantUriPermissions="true">
24           <meta-data
25               android:name="android.support.FILE_PROVIDER_PATHS"
26               android:resource="@xml/file_paths"></meta-data>
27       </provider>
```

第 03～05 行设置摄像头和外部存储读写权限。第 19～27 行设置 FileProvider 的相关标签属性，android:authorities="com.vt.c0801.fileprovider"设置授权字符串，android:resource="@xml/file_paths"指定设置权限文件夹的文件。

6．主界面的 Activity

```
/java/com/vt/c0801/MainActivity.java
16   public class MainActivity extends AppCompatActivity implements View.OnClickListener {
17       private static final int REQUEST_CODE_TAKE_PHOTO_DEFAULT = 0;
18       private static final int REQUEST_CODE_TAKE_PHOTO_CUSTOM = 1;
19       private static final int REQUEST_CODE_SELECT = 2;
20       private static final int REQUEST_CODE_CROP = 3;
21       private Context mContext = this;
```

```java
22      private File mPhotoFile;
23      //需要请求的权限
24      private static final String[] PERMISSIONS = {
25              Manifest.permission.CAMERA,
26              Manifest.permission.READ_EXTERNAL_STORAGE,
27              Manifest.permission.WRITE_EXTERNAL_STORAGE
28      };
29
30      @Override
31      protected void onCreate(Bundle savedInstanceState) {
32          super.onCreate(savedInstanceState);
33          setContentView(R.layout.activity_main);
34          //初始化
35          this.setTitle("C0801:拍照、选取和显示图片");
36          findViewById(R.id.button_take_photo_default).setOnClickListener(this);
37          findViewById(R.id.button_take_photo_custom).setOnClickListener(this);
38          findViewById(R.id.button_select_photo).setOnClickListener(this);
39          //判断是否已经取得权限
40          if (!Permissions.hasPermissionsGranted(mContext, PERMISSIONS)) {
41              //请求权限
42              Permissions.requestPermissions(mContext, PERMISSIONS);
43          }
44      }
45      //请求权限的回调
46      @Override
47      public void onRequestPermissionsResult(int requestCode, String[] permissions, int[] grantResults) {
48          switch (requestCode) {
49              case Permissions.REQUEST_PERMISSIONS:
50                  if (!Permissions.hasPermissionsGranted(mContext, PERMISSIONS)) {
51                      Permissions.requestPermissions(mContext, PERMISSIONS);
52                  }
53                  break;
54          }
55      }
56      //单击事件
57      @Override
58      public void onClick(View v) {
59          switch (v.getId()) {
60              case R.id.button_take_photo_default: {
61                  takePhotoDefault();
62                  break;
63              }
64              case R.id.button_take_photo_custom: {
65                  takePhotoCustom();
66                  break;
67              }
68              case R.id.button_select_photo: {
69                  selectPhoto();
70                  break;
```

```java
71              }
72          }
73      }
74      //系统相机拍照，使用默认保存路径
75      private void takePhotoDefault() {
76          //调用系统相机拍照
77          Intent intent = new Intent(MediaStore.ACTION_IMAGE_CAPTURE);
78          startActivityForResult(intent, REQUEST_CODE_TAKE_PHOTO_DEFAULT);
79      }
80      //系统相机拍照，照片保存到指定路径
81      private void takePhotoCustom() {
82          //创建文件
83          mPhotoFile = Util.createFile(mContext.getExternalMediaDirs()[0].getAbsolutePath(),
                    mContext.getResources().getString(R.string.app_name), "jpg");
84          if (mPhotoFile != null) {
85              //获取"com.vt.c0801.fileprovider"授权路径下的mPhotoFile文件的URI
86              Uri photoURI = FileProvider.getUriForFile(this, "com.vt.c0801.fileprovider", mPhotoFile);
87              //调用系统相机拍照
88              Intent intent = new Intent(MediaStore.ACTION_IMAGE_CAPTURE);
89              intent.putExtra(MediaStore.EXTRA_OUTPUT, photoURI);//设置保存文件的URI地址
90              startActivityForResult(intent, REQUEST_CODE_TAKE_PHOTO_CUSTOM);
91          }
92      }
93      //调用系统相册选择图片
94      public void selectPhoto() {
95          Intent intent = new Intent(Intent.ACTION_PICK,
                    android.provider.MediaStore.Images.Media.EXTERNAL_CONTENT_URI);
96          intent.setType("image/*");
97          startActivityForResult(intent, REQUEST_CODE_SELECT);
98      }
99      // Activity回调
100     @Override
101     protected void onActivityResult(int requestCode, int resultCode, Intent intent) {
102         if (requestCode == REQUEST_CODE_TAKE_PHOTO_DEFAULT && resultCode == RESULT_OK) {
103             //从data中取出传递回来缩略图的信息，图片质量差，适合传递小图片
104             Bundle bundle = intent.getExtras();
105             Bitmap bitmap = bundle.getParcelable("data");
106             //显示图片
107             ((ImageView) findViewById(R.id.image_view)).setImageBitmap(bitmap);
108         } else if (requestCode == REQUEST_CODE_TAKE_PHOTO_CUSTOM && resultCode == RESULT_OK) {
109             //显示图片
110             ((ImageView) findViewById(R.id.image_view)).setImageURI(Uri.fromFile(mPhotoFile));
111             //扫描更新相册
112             Util.showInAlbum(mContext, mPhotoFile.getAbsolutePath());
113         } else if (requestCode == REQUEST_CODE_SELECT && resultCode == RESULT_OK) {
114             //新建保存裁剪图片的文件
115             mPhotoFile = Util.createFile(mContext.getExternalMediaDirs()[0].getAbsolutePath(),
                        mContext.getResources().getString(R.string.app_name), "jpg");
116             //裁剪图片
117             Intent cropIntent = new Intent("com.android.camera.action.CROP");
```

```
118              cropIntent.setDataAndType(intent.getData(), "image/*");
119              cropIntent.putExtra("crop", "true");
120              //设置裁剪的比例
121              cropIntent.putExtra("aspectX", 1);
122              cropIntent.putExtra("aspectY", 1);
123              //设置裁剪后保存图片的尺寸
124              cropIntent.putExtra("outputX", 500);
125              cropIntent.putExtra("outputY", 500);
126              //设置保存图片的文件格式
127              cropIntent.putExtra("outputFormat", Bitmap.CompressFormat.JPEG.toString());
128              //设置保存图片的文件路径
129              cropIntent.putExtra(MediaStore.EXTRA_OUTPUT, Uri.parse("file://" +
                     mPhotoFile.getAbsolutePath()));
130              cropIntent.putExtra("return-data", false);
131              startActivityForResult(cropIntent, REQUEST_CODE_CROP);
132              //扫描更新相册
133              Util.showInAlbum(mContext, mPhotoFile.getAbsolutePath());
134          }else if (requestCode == REQUEST_CODE_CROP && resultCode == RESULT_OK) {
135              //显示图片
136              ((ImageView) findViewById(R.id.image_view)).setImageURI(Uri.fromFile(mPhotoFile));
137          }
138      }
139  }
```

第 40~43 行判断是否已经获取授权，如果首次运行时没有获取所有授权，或请求的权限在应用设置中设置为"询问"，则打开系统授权对话框。第 47~55 行处理系统授权对话框对权限选择结果的回调，如果拒绝权限申请，则打开自定义对话框，提示用户手动开启权限。第 77 行 MediaStore.ACTION_IMAGE_CAPTURE 表示拍照的行为，拍照时可以选择系统相机或第三方相机 App。第 86 行生成的 URI 格式为"content://com.vt.c0801.fileprovider/cache_images/+文件名称"，如 content://com.vt.c0801.fileprovider/cache_images/IMG_20200213_010345.jpg。第 95 行 Intent.ACTION_PICK 表示选择的行为，android.provider.MediaStore.Images.Media.EXTERNAL_CONTENT_URI 表示提供图片的 URI。第 115 行 mContext.getExternalMediaDirs()[0].getAbsolutePath() 获取系统媒体文件夹的路径，mContext.getResources().getString(R.string.app_name) 获取 App 的名称。第 117~127 行在选择图片后调用系统的图片剪裁功能，并将剪裁后图片保存。第 129 行设置剪裁文件保存的文件路径，此处还不支持 FileProvider，而使用 Uri.parse() 方法通过"file://"形式生成 URI。

8.1.2 实例工程：录制、选取和播放视频

本实例演示了使用系统相机 App 录制视频及使用系统相册选取视频，使用 VideoView 控件播放视频，录制的视频会自动保存到系统相册中。使用系统相机录制视频同样需要获取相应的权限，单击"录制视频"按钮调用系统相机可以录制 15s 内的视频，单击"选取视频"按钮调用系统相册，选取视频后在 VideoView 控件中播放，单击 VideoView 控件可以显示播放控制器（如图 8-3 所示）。

图 8-3 运行效果

1. 打开基础工程

打开"基础工程"文件夹中的"C0802"工程，该工程已经包含 MainActivity、Permissions 及布局的文件。Permissions 类与"C0801"工程的 Permissions 类相同，是可以直接复制而重复使用的。activity_layout.xml 布局文件包含一个 VideoView 控件用于播放视频。

2. 主界面的 Activity

```
/java/com/vt/c0802/MainActivity.java
49   //调用系统相机录制视频
50   public void recordVideo() {
51       Intent intent = new Intent(MediaStore.ACTION_VIDEO_CAPTURE);
52       intent.putExtra(MediaStore.EXTRA_VIDEO_QUALITY, 1);//设置视频质量（1为高画质）
53       intent.putExtra(MediaStore.EXTRA_DURATION_LIMIT, 15);//设置最长时间为15s
54       startActivityForResult(intent, REQUEST_CODE_RECORD);
55   }
56   //调用系统相册选取视频
57   public void selectVideo() {
58       Intent intent = new Intent(Intent.ACTION_PICK, android.provider.MediaStore.Video.Media.EXTERNAL_CONTENT_URI);
59       startActivityForResult(intent, REQUEST_CODE_SELECT);
60   }
74   //回调播放录制的视频
75   Override
76   protected void onActivityResult(int requestCode, int resultCode, Intent intent) {
77       if ((requestCode == REQUEST_CODE_RECORD || requestCode == REQUEST_CODE_SELECT) && resultCode == RESULT_OK) {
78           Uri videoUri = intent.getData();
79           final VideoView videoView = findViewById(R.id.video_view);
80           videoView.setVideoURI(videoUri);
81           videoView.setMediaController(new MediaController(this));
```

```
82          videoView.setOnPreparedListener(new MediaPlayer.OnPreparedListener() {
83              @Override
84              public void onPrepared(MediaPlayer mp) {
85                  videoView.start();
86              }
87          });
88      }
89  }
```

第 51 行 MediaStore.ACTION_VIDEO_CAPTURE 是录制视频的行为。第 52 行设置录制视频的画质，1 代表高画质，0 代表低画质。第 53 行设置录制视频的最长时间，时间单位为 s，录制界面会出现倒计时的数字提示。第 58 行 Intent.ACTION_PICK 表示选取的行为，android.provider.MediaStore.Video.Media.EXTERNAL_CONTENT_URI 表示提供视频的 URI。第 78 行 intent.getData()获取选取或录制的视频 URI。第 80 行设置 VideoView 控件播放的视频 URI。第 81 行新建一个媒体控制器作为 VideoView 控件的播放控制器，播放视频时单击视频会出现控制器，再次单击隐藏控制器。第 82~87 行设置视频准备完成事件的监听器，当视频准备完成后播放视频。

8.2 拍摄照片和录制视频

调用系统相机虽然能够拍摄照片和录制视频，但是所提供的功能有限且灵活性差。使用 API 提供的类直接控制摄像头更加灵活，可以编写自定义功能的相机。目前 Android 提供的摄像头 API 共有 3 个版本，分别为 Camera 类、Camera2 类和 CameraX 类。使用这 3 个类都需要在 AndroidManifest.xml 文件中设置 Manifest.permission.camera 权限，官方推荐使用 Camera2 类。因为 Camera 类在 API level 21 就过时了，CameraX 类目前为 alpha 版本。CameraX 类是为了解决 Camera2 类的复杂性和硬件设备的兼容性问题，很容易实现人像、HDR、夜间模式和美颜等效果，是未来使用摄像头的 API。

8.2.1 Camera2 类

Camera2 类是 android.hardware.camera2 包提供的涉及摄像头管理和使用的类集合。Camera2 类提供的功能比 Camera 类要丰富得多，增加了使用的复杂度。Camera2 类将摄像头作为管道，该管道接收输入请求以捕获单个帧，每个请求捕获单个图像，然后输出一个捕获结果元数据包及该请求的一组输出图像缓冲区。请求需要按顺序处理，多个请求可以一次运行。

CameraManager 类（android.hardware.camera2.CameraManager）用于检测、描述、查询和连接可用的摄像头（如表 8-1 所示）。需要使用 Context.getSystemService(Context.CAMERA_SERVICE)或 Context.getSystemService(CameraManager.class)方法获取该实例。

表 8-1 CameraManager 类的常用方法

类型和修饰符	方 法
CameraCharacteristics	getCameraCharacteristics(String cameraId) 查询摄像头的功能

续表

类型和修饰符	方法
String[]	getCameraIdList() 获取可用的摄像头 id 列表
void	openCamera(String cameraId, CameraDevice.StateCallback callback, Handler handler) 打开摄像头
void	registerAvailabilityCallback(CameraManager.AvailabilityCallback callback, Handler handler) 注册摄像头优先级和可用性的回调
void	registerTorchCallback(CameraManager.TorchCallback callback, Handler handler) 注册闪光灯的回调
void	setTorchMode(String cameraId, boolean enabled) 设置闪光灯的闪光灯模式
void	unregisterAvailabilityCallback(CameraManager.AvailabilityCallback callback) 注销摄像头优先级和可用性的回调
void	unregisterTorchCallback(CameraManager.TorchCallback callback) 注销闪光灯的回调

CameraManager.AvailabilityCallback 类（android.hardware.camera2.CameraManager.AvailabilityCallback）用于处理摄像头的优先级改变或可用性改变的回调（如表 8-2 所示），需要使用 CameraManager 类的 registerAvailabilityCallback(CameraManager.AvailabilityCallback, Handler)方法和 unregisterAvailabilityCallback(CameraManager.AvailabilityCallback)方法进行注册和注销。

表 8-2　CameraManager.AvailabilityCallback 类的常用方法

类型和修饰符	方法
	CameraManager.AvailabilityCallback() 构造方法
void	onCameraAccessPrioritiesChanged() 摄像头访问优先级改变时的回调方法
void	onCameraAvailable(String cameraId) 摄像头可用时的回调方法
void	onCameraUnavailable(String cameraId) 摄像头不可用时的回调方法

CameraManager.TorchCallback 类（android.hardware.camera2.CameraManager.TorchCallback）用于闪光灯模式可用性改变时或不可用时的回调（如表 8-3 所示），需要使用 CameraManager 类的 registerTorchCallback(CameraManager.TorchCallback, Handler)方法和 unregisterTorchCallback (CameraManager.TorchCallback)方法进行注册和注销。

表 8-3　CameraManager.TorchCallback 类的常用方法

类型和修饰符	方法
	CameraManager.TorchCallback() 构造方法
void	onTorchModeChanged(String cameraId, boolean enabled) 摄像头闪光灯模式改变时的回调
void	onTorchModeUnavailable(String cameraId) 摄像头闪光灯模式不可用时的回调

CameraDevice 类（android.hardware.camera2.CameraDevice）用于摄像头对图像捕捉的请求和会话进行处理（如表 8-4 所示）。

表 8-4　CameraDevice 类的常用方法

类型和修饰符	方　　法
abstract void	close() 关闭与摄像头的连接
CaptureRequest.Builder	createCaptureRequest(int templateType, Set<String> physicalCameraIdSet) 创建捕捉请求。templateType 参数表示模板类型，使用以下常量表示。 ● CameraDevice.TEMPLATE_MANUAL：手动模式 ● CameraDevice.TEMPLATE_PREVIEW：摄像头预览模式 ● CameraDevice.TEMPLATE_RECORD：视频录制模式 ● CameraDevice.TEMPLATE_STILL_CAPTURE：拍照模式 ● CameraDevice.TEMPLATE_VIDEO_SNAPSHOT：录制视频时的拍照模式 ● CameraDevice.TEMPLATE_ZERO_SHUTTER_LAG：零快门滞后仍然捕获模式
abstract CaptureRequest.Builder	createCaptureRequest(int templateType) 创建捕捉请求。templateType 参数与 createCaptureRequest(int, Set<String>)的 templateType 参数相同
abstract void	createCaptureSession(List<Surface> outputs, CameraCaptureSession.StateCallback callback, Handler handler) 创建捕捉会话
abstract CaptureRequest.Builder	createReprocessCaptureRequest(TotalCaptureResult inputResult) 创建重新处理的捕捉请求
abstract void	createReprocessableCaptureSession(InputConfiguration inputConfig, List<Surface> outputs, CameraCaptureSession.StateCallback callback, Handler handler) 创建重新处理的捕捉会话
abstract String	getId() 获取摄像头的 id

CameraDevice.StateCallback 类（android.hardware.camera2.CameraDevice.StateCallback）用于对摄像头的开启、关闭、断开连接和报错时的回调（如表 8-5 所示）。使用 Camera-Manager.openCamera(String, CameraDevice.StateCallback, Handler)方法来打开摄像头后，才能处理回调。

表 8-5　CameraDevice.StateCallback 类的常用方法

类型和修饰符	方　　法
void	onClosed(CameraDevice camera) 关闭摄像头时调用该方法
abstract void	onDisconnected(CameraDevice camera) 摄像头断开连接不再可用时调用该方法
abstract void	onError(CameraDevice camera, int error) 摄像头遇到错误时调用该方法。error 参数表示发生的错误类型，使用以下常量表示。 ● CameraDevice.StateCallback.ERROR_CAMERA_DEVICE：摄像头硬件错误 ● CameraDevice.StateCallback.ERROR_CAMERA_DISABLED：摄像头不可用 ● CameraDevice.StateCallback.ERROR_CAMERA_IN_USE：摄像头正在使用中 ● CameraDevice.StateCallback.ERROR_CAMERA_SERVICE：摄像头服务的错误 ● CameraDevice.StateCallback.ERROR_MAX_CAMERAS_IN_USE：到达最大使用数量
abstract void	onOpened(CameraDevice camera) 摄像头被打开后调用该方法

CameraCaptureSession 类（android.hardware.camera2.CameraCaptureSession）用于设置和管理捕捉图像会话（如表 8-6 所示）。使用 CameraDevice 的 createCaptureSession()方法或 createReprocessableCaptureSession()方法创建的捕捉图像会话属于异步操作，因为需要配置摄像头的内部管道，并分配内存缓冲区，以便将捕捉的图像发送到所需的目标。捕捉图像会话创建后就会处于活动状态，并激活 CameraCaptureSession.StateCallback 的回调方法。

表 8-6　CameraCaptureSession 类的常用方法

类型和修饰符	方法
	CameraCaptureSession() 构造方法
abstract void	abortCaptures() 丢弃当前挂起和正在进行的所有捕捉图像会话
abstract int	capture(CaptureRequest request, CameraCaptureSession.CaptureCallback listener, Handler handler) 提交捕捉图像请求的会话
abstract int	captureBurst(List\<CaptureRequest\> requests, CameraCaptureSession.CaptureCallback listener, Handler handler) 提交捕捉图像序列请求的会话，序列按照顺序执行
int	captureBurstRequests(List\<CaptureRequest\> requests, Executor executor, CameraCaptureSession.CaptureCallback listener) 提交捕捉图像序列请求的会话
int	captureSingleRequest(CaptureRequest request, Executor executor, CameraCaptureSession.CaptureCallback listener) 提交单个捕捉图像请求的会话
abstract void	close() 异步关闭此捕捉图像会话
abstract CameraDevice	getDevice() 获取为此会话创建的摄像头设备
abstract Surface	getInputSurface() 获取输入捕捉图像会话关联的 Surface
abstract boolean	isReprocessable() 判断是否可以重新提交捕捉请求的会话
abstract void	prepare(Surface surface) 准备捕捉，为用于捕捉图像输出的 Surface 预先分配缓冲区
abstract int	setRepeatingBurst(List\<CaptureRequest\> requests, CameraCaptureSession.CaptureCallback listener, Handler handler) 设置重复捕捉图像的序列
int	setRepeatingBurstRequests(List\<CaptureRequest\> requests, Executor executor, CameraCaptureSession.CaptureCallback listener) 设置重复捕捉图像的序列请求
abstract int	setRepeatingRequest(CaptureRequest request, CameraCaptureSession.CaptureCallback listener, Handler handler) 设置重复捕捉图像的请求
int	setSingleRepeatingRequest(CaptureRequest request, Executor executor, CameraCaptureSession.CaptureCallback listener) 设置单个重复捕捉图像的请求

续表

类型和修饰符	方法
abstract void	stopRepeating() 取消使用 setRepeatingRequest(CaptureRequest, CameraCaptureSession.CaptureCallback, Handler)或 setRepeatingRequest(CaptureRequest, CameraCaptureSession.CaptureCallback, Handler)方法设置的重复捕捉图像请求
void	updateOutputConfiguration(OutputConfiguration config) 更新最终确定的输出配置

CameraCaptureSession.CaptureCallback 类（android.hardware.camera2.CameraCaptureSession.CaptureCallback）用于跟踪提交的 CaptureRequest 对象（如表 8-7 所示）。

表 8-7 CameraCaptureSession.CaptureCallback 类的常用方法

类型和修饰符	方法
	CameraCaptureSession.CaptureCallback() 构造方法
void	onCaptureBufferLost(CameraCaptureSession session, CaptureRequest request, Surface target, long frameNumber) 当无法将单个捕捉图像缓冲发送到 surface 时，调用此方法
void	onCaptureCompleted(CameraCaptureSession session, CaptureRequest request, TotalCaptureResult result) 当捕捉图像完成后且所有结果元数据都可用时，调用此方法
void	onCaptureFailed(CameraCaptureSession session, CaptureRequest request, CaptureFailure failure) 当捕捉图像失败无法生成 CaptureResult 时，调用此方法
void	onCaptureProgressed(CameraCaptureSession session, CaptureRequest request, CaptureResult partialResult) 当捕捉图像部分进度完成后，调用此方法，可以使用已经捕捉的图像结果
void	onCaptureSequenceAborted(CameraCaptureSession session, int sequenceId) 当捕捉图像序列被中止后，调用此方法
void	onCaptureSequenceCompleted(CameraCaptureSession session, int sequenceId, long frameNumber) 当捕捉图像序列完成后，调用此方法
void	onCaptureStarted(CameraCaptureSession session, CaptureRequest request, long timestamp, long frameNumber) 当捕捉图像请求开始曝光照片或开始处理输入的图像后，调用此方法

CameraCaptureSession.StateCallback 类（android.hardware.camera2.CameraCaptureSession.StateCallback）用于接收捕捉图像会话状态的改变（如表 8-8 所示）。

表 8-8 CameraCaptureSession.StateCallback 类的常用方法

类型和修饰符	方法
	CameraCaptureSession.StateCallback() 构造方法
void	onActive(CameraCaptureSession session) 当会话开始主动处理捕捉图像请求时，调用此方法
void	onCaptureQueueEmpty(CameraCaptureSession session) 当捕捉图像队列为空并准备接收下一个请求时，调用此方法
void	onClosed(CameraCaptureSession session) 当捕捉图像会话关闭后，调用此方法
abstract void	onConfigureFailed(CameraCaptureSession session) 当配置失败后，调用此方法

续表

类型和修饰符	方法
abstract void	onConfigured(CameraCaptureSession session) 当配置成功后，调用此方法
void	onReady(CameraCaptureSession session) 当会话不再有任何请求时，调用此方法
void	onSurfacePrepared(CameraCaptureSession session, Surface surface) 当输出 Surface 的缓冲区预分配已完成后，调用此方法

CaptureRequest 类（android.hardware.camera2.CaptureRequest）用来配置捕捉单个图像（如表 8-9 所示）。CaptureRequest 对象调用 CameraCaptureSession 的 capture(CaptureRequest, CameraCaptureSession.CaptureCallback, Handler) 或 setRepeatingRequest(CaptureRequest, CameraCaptureSession.CaptureCallback, Handler) 方法从摄像头中捕捉拍摄的图像数据。

每个请求可为摄像头指定一个 Surface，将捕捉的图像数据发送到该 Surface。当请求提交到会话时，请求中使用的所有 Surface 都必须包含在最后一次调用 CameraDevice.createCaptureSession(List<Surface>, CameraCaptureSession.StateCallback, Handler)方法的 Surface 列表参数中。例如，一个请求用于低分辨率图像的预览，另一个请求用于高分辨率图像的捕捉。

重新处理捕捉请求允许将先前从相机设备捕捉的图像发送回设备进一步处理，实现双缓冲的功能。除了通过 CameraDevice.createCaptureSession()方法创建的捕捉图像会话，还可以通过 CameraDevice.createReprocessableCaptureSession()方法创建可重新处理的捕捉图像会话，能够在提交常规捕捉图像请求外的重新处理捕捉图像请求。重新处理捕捉图像请求从会话的输入 Surface 获取下一个可用缓冲区，再次通过摄像头的管道将其作为 Surface 的缓冲区发送，此时没有为重新处理请求捕捉新的图像数据。

表 8-9　CaptureRequest 类的常用方法

类型和修饰符	方法
<T> T	get(Key<T> key) 获取键值
List<Key<?>>	getKeys() 获取此映射中包含的键列表
Object	getTag() 获取标签
boolean	isReprocess() 判断是否重新处理捕捉图像请求

CaptureRequest.Builder 类（android.hardware.camera2.CaptureRequest.Builder）用于构建捕捉图像请求（如表 8-10 所示）。可以使用 CameraDeviced.createCaptureRequest(int)方法获得构建器实例，该方法将请求字段初始化为在 CameraDevice 定义的一个模板。

表 8-10　CaptureRequest.Builder 类的常用方法

类型和修饰符	方法
void	addTarget(Surface outputTarget) 添加捕捉图像输出的 Surface，添加的 Surface 必须包含在 CameraDevice.createCaptureSession(List<Surface>,CameraCaptureSession.StateCallback, Handler)方法的参数中

续表

类型和修饰符	方 法
CaptureRequest	build() 构建捕捉图像请求
<T> T	get(Key<T> key) 获取键值
void	removeTarget(Surface outputTarget) 移除捕捉图像输出的 Surface
<T> void	set(Key<T> key, T value) 设置键值
void	setTag(Object tag) 获取标签

CameraCharacteristics 类（android.hardware.camera2.CameraCharacteristics）用于描述摄像头设备的属性（如表 8-11 所示），使用 CameraManager.getCameraCharacteristics(String) 方法获取单个摄像头的 CameraCharacteristics 对象。

表 8-11　CameraCharacteristics 类的常用方法

类型和修饰符	方 法
<T> T	get(Key<T> key) 获取键值
List<Key<?>>	getAvailableCaptureRequestKeys() 获取有效的捕捉图像请求的键列表
List<Key<?>>	getAvailablePhysicalCameraRequestKeys() 获取有效的物理摄像头请求的键列表，物理摄像头可以虚拟多个逻辑摄像头
List<Key<?>>	getAvailableSessionKeys() 获取有效的会话键列表
List<Key<?>>	getKeys() 获取键列表
List<Key<?>>	getKeysNeedingPermission() 获取需要 Manifest.permission.CAMERA 权限支持的所有键
Set<String>	getPhysicalCameraIds() 获取物理摄像头 id

CameraMetadata 类（android.hardware.camera2.CameraMetadata<Tkey>）用于保存摄像头特性的元数据，只提供了一个 getKeys()方法用于获取键列表，但是提供了非常多的表示摄像头特性的常量（如表 8-12 所示）。

表 8-12　CameraMetadata 类的部分常量

类型	常　　量	说　　明
int	CONTROL_AE_ANTIBANDING_MODE_AUTO	摄像头将自动调整防止频闪
int	CONTROL_AE_ANTIBANDING_MODE_OFF	摄像头将不会调整曝光时间以免出现频闪
int	CONTROL_AE_MODE_OFF	摄像头关闭自动曝光模式
int	CONTROL_AE_MODE_ON	摄像头启动自动曝光模式
int	CONTROL_AE_MODE_ON_ALWAYS_FLASH	拍摄照片始终使用闪光灯模式
int	CONTROL_AE_MODE_ON_AUTO_FLASH	弱光情况下拍摄照片使用闪光灯模式
int	CONTROL_AF_MODE_AUTO	自动对焦模式

续表

类型	常量	说明
int	CONTROL_AF_MODE_CONTINUOUS_PICTURE	拍摄照片连续对焦模式
int	CONTROL_AF_MODE_CONTINUOUS_VIDEO	录制视频持续对焦模式
int	CONTROL_AF_MODE_MacRO	微距对焦模式
int	CONTROL_AF_MODE_OFF	关闭自动对焦模式
int	CONTROL_AWB_MODE_AUTO	自动白平衡模式
int	CONTROL_AWB_MODE_CLOUDY_DAYLIGHT	日光多云白平衡模式
int	CONTROL_AWB_MODE_DAYLIGHT	日光白平衡模式
int	CONTROL_AWB_MODE_FLUORESCENT	荧光灯白平衡模式
int	CONTROL_AWB_MODE_INCANDESCENT	白炽灯白平衡模式
int	CONTROL_AWB_MODE_OFF	关闭自动白平衡模式
int	CONTROL_AWB_MODE_SHADE	阴影白平衡模式
int	CONTROL_AWB_MODE_TWILIGHT	暮光白平衡模式
int	CONTROL_EFFECT_MODE_MONO	单色效果模式
int	CONTROL_EFFECT_MODE_NEGATIVE	负片效果模式
int	CONTROL_EFFECT_MODE_OFF	关闭色彩效果模式
int	CONTROL_EFFECT_MODE_POSTERIZE	色块效果模式
int	CONTROL_MODE_AUTO	全自动模式
int	CONTROL_MODE_OFF	关闭全自动模式
int	CONTROL_SCENE_MODE_BARCODE	条形码场景模式
int	CONTROL_SCENE_MODE_DISABLED	关闭场景模式
int	CONTROL_SCENE_MODE_FIREWORKS	烟花场景模式
int	CONTROL_SCENE_MODE_HDR	高动态范围（HDR）模式
int	CONTROL_SCENE_MODE_LANDSCAPE	风景模式
int	CONTROL_SCENE_MODE_NIGHT	夜景模式
int	CONTROL_SCENE_MODE_NIGHT_PORTRAIT	夜景人像模式
int	CONTROL_SCENE_MODE_PORTRAIT	肖像模式
int	CONTROL_SCENE_MODE_SNOW	雪景模式
int	CONTROL_SCENE_MODE_SPORTS	运动场景模式
int	CONTROL_VIDEO_STABILIZATION_MODE_OFF	关闭视频数码稳定模式
int	CONTROL_VIDEO_STABILIZATION_MODE_ON	开启视频数码稳定模式
int	FLASH_MODE_OFF	关闭闪光灯模式
int	FLASH_MODE_SINGLE	单次闪光灯模式
int	FLASH_MODE_TORCH	闪光灯常亮模式
int	LENS_FACING_BACK	与屏幕朝向方向相反的摄像头，即后置摄像头
int	LENS_FACING_EXTERNAL	外置摄像头
int	LENS_FACING_FRONT	与屏幕朝向方向相同的摄像头，即前置摄像头

8.2.2 ImageReader 类

Camera 类通过 Camera.takePicture()方法的参数——Camera.PictureCallback 实例回调获取拍摄的照片数据。而 Camera2 类则需要通过 ImageReader 类读取拍摄时的照片数据。

ImageReader 类（android.media.ImageReader）用于读取 Surface 对象存储的图像数据（如表 8-13 所示）。调用 ImageReader.acquireLatestImage()方法或 ImageReader.acquireNextImage()

方法读取 Surface 对象提供的图像队列。由于内存限制，如果不能以相同的速率获取和释放图像，那么 Surface 对象将停止提供或删除图像。

表 8-13　ImageReader 类的常用方法

类型和修饰符	方　　　　法
Image	acquireLatestImage() 从 ImageReader 的队列中获取最新的图像，删除较旧的图像
Image	acquireNextImage() 从 ImageReader 的队列中获取下一个图像
void	close() 关闭并释放资源
int	getHeight() 获取图像高度
int	getImageFormat() 获取图像格式
int	getMaxImages() 获取可获取的最大图像数量
Surface	getSurface() 获取读取图像的 Surface
int	getWidth() 获取图像宽度
static ImageReader	newInstance(int width, int height, int format, int maxImages) 创建 ImageReader 的新实例
void	setOnImageAvailableListener(ImageReader.OnImageAvailableListener listener, Handler handler) 设置从 ImageReader 获得新图像时调用的监听器

8.2.3　MediaRecorder 类

MediaRecorder 类（android.media.MediaRecorder）用于录制视频参数的设置，以及录制、暂停等操作（如表 8-14 所示），录制的视频中可以包含音频，同时适用于 Camera 类和 Camera2 类录制视频。也可以录制摄像头或其他视频源捕捉的视频。

表 8-14　MediaRecorder 类的常用方法

类型和修饰符	方　　　　法
	MediaRecorder() 构造方法
Surface	getSurface() 获取录制视频源的 Surface
void	pause() 暂停录制
void	prepare() 准备开始捕捉和编码数据
void	registerAudioRecordingCallback(Executor executor, AudioManager.AudioRecordingCallback cb) 注册音频录制的回调
void	release() 释放资源
void	reset() 重置
void	resume() 恢复录制

续表

类型和修饰符	方法
void	setAudioEncoder(int audio_encoder) 设置音频编码器。audio_encoder 的常量如下。 ● MediaRecorder.AudioEncoder.AAC：高级音频编码（Advanced Audio Coding），与 mp3 相比，AAC 格式的音质更佳，文件更小 ● MediaRecorder.AudioEncoder.AAC_ELD：低延迟的 AAC ● MediaRecorder.AudioEncoder.AMR_NB：窄频 AMR ● MediaRecorder.AudioEncoder.AMR_WB：宽频 AMR ● MediaRecorder.AudioEncoder.DEFAULT：默认编码 ● MediaRecorder.AudioEncoder.HE_AAC：高保真 AAC ● MediaRecorder.AudioEncoder.OPUS：Opus 编码器。低码率下 Opus 完胜 HE AAC，中码率下与码率高出 30%左右的 AAC 格式相当，高码率下更接近原始音频 ● MediaRecorder.AudioEncoder.VORBIS：Ogg Vorbis 编码器，不用重新编码即可调节文件的位速率
void	setAudioEncodingBitRate(int bitRate) 设置音频编码比特率
void	setAudioSamplingRate(int samplingRate) 设置音频采样率
void	setAudioSource(int audioSource) 设置音源
void	setCaptureRate(double fps) 设置捕捉的帧速率
void	setInputSurface(Surface surface) 设置录制视频源的 Surface
void	setLocation(float latitude, float longitude) 设置 GPS 坐标
void	setMaxDuration(int max_duration_ms) 设置录制的最长时间（ms）
void	setMaxFileSize(long max_filesize_bytes) 设置最大文件尺寸（Byte）
void	setOnErrorListener(MediaRecorder.OnErrorListener l) 设置捕捉错误的监听器
void	setOnInfoListener(MediaRecorder.OnInfoListener listener) 设置信息事件的监听器。监听的信息事件如下。 ● MediaRecorder.MEDIA_RECORDER_INFO_UNKNOWN：未知信息 ● MediaRecorder.MEDIA_RECORDER_INFO_MAX_DURATION_REACHED：达到最长录制时间 ● MediaRecorder.MEDIA_RECORDER_INFO_MAX_FILESIZE_REACHED：录制的输出文件已经达到最大文件尺寸
void	setOrientationHint(int degrees) 设置输出视频回放的方向提示
void	setOutputFile(FileDescriptor fd) 设置录制的输出文件
void	setOutputFile(String path) 设置录制的输出文件路径
void	setOutputFile(File file) 设置录制的输出文件
void	setOutputFormat(int output_format) 设置录制的输出格式
boolean	setPreferredMicrophoneFieldDimension(float zoom) 设置麦克风声音大小的缩放
void	setPreviewDisplay(Surface sv) 设置预览视频显示的 Surface

续表

类型和修饰符	方　法
void	setVideoEncoder(int video_encoder) 设置录制的视频编码。video_encoder 的常量如下： ● MediaRecorder.VideoEncoder.DEFAULT ● MediaRecorder.VideoEncoder.H263 ● MediaRecorder.VideoEncoder.H264 ● MediaRecorder.VideoEncoder.HEVC ● MediaRecorder.VideoEncoder.MPEG_4_SP ● MediaRecorder.VideoEncoder.VP8
void	setVideoEncodingBitRate(int bitRate) 设置录制的视频编码比特率
void	setVideoFrameRate(int rate) 设置录制的视频帧速率
void	setVideoSize(int width, int height) 设置录制的视频尺寸
void	setVideoSource(int video_source) 设置录制的视频源
void	start() 开始录制视频，捕捉视频进行编码并写入输出文件
void	stop() 停止录制
void	unregisterAudioRecordingCallback(AudioManager.AudioRecordingCallback cb) 注销声音录制的回调

8.2.4　实例工程：使用 Camera2 类拍摄照片

本实例演示了使用 Camera2 类开发第三方拍照 App 的方法，单击"拍照"按钮后拍摄照片并显示照片（如图 8-4 所示）。在其他 App 调用系统相机时，该 App 可以作为备选项供用户选择。

图 8-4　运行效果

1. 打开基础工程

打开"基础工程"文件夹中的"C0805"工程，该工程已经包含 MainActivity、PreviewActivity、Permissions、PreviewTextureView、Util 类及布局的文件，在 AndroidManifest.xml 文件中还添加了拍摄和存储照片所需的权限及隐性启动的行为。

2. Util 类

打开 Util 类，添加两个可重用的静态方法 imageToBytes()和 chooseOptimalSize()，以及一个可重用的内部类 CompareSizesByArea。

```
/java/com/vt/c0805/Util.java
106    //bitmap 转 byte[]
107    public static byte[] imageToBytes(Image image){
108        ByteBuffer buffer = image.getPlanes()[0].getBuffer();
109        byte[] bytes = new byte[buffer.remaining()];
110        buffer.get(bytes);
111        return bytes;
112    }
113    //获取最佳尺寸
114    public static Size chooseOptimalSize(Context context, Size[] choices, int previewWidth, int previewHeight, int maxPreviewWidth, int maxPreviewHeight, Size aspectRatio) {
115        Point displaySize = new Point();
116        //获取屏幕尺寸
117        ((Activity) context).getWindowManager().getDefaultDisplay().getSize(displaySize);
118        int textureViewWidth = previewWidth;
119        int textureViewHeight = previewHeight;
120        int maxWidth = displaySize.x;
121        int maxHeight = displaySize.y;
122        if (maxWidth > maxPreviewWidth) { maxWidth = maxPreviewWidth; }
123        if (maxHeight > maxPreviewHeight) { maxHeight = maxPreviewHeight; }
124        List<Size> bigEnough = new ArrayList<>();//用于保存支持的不小于预览图像的尺寸
125        List<Size> notBigEnough = new ArrayList<>();//用于保存支持的小于预览图像的尺寸
126        int w = aspectRatio.getWidth();
127        int h = aspectRatio.getHeight();
128        //将设备支持的尺寸分类保存
129        for (Size option : choices) {
130            if (option.getWidth() <= maxWidth && option.getHeight() <= maxHeight && option.getHeight() == option.getWidth() * h / w) {
131                if (option.getWidth() >= textureViewWidth &&
132                        option.getHeight() >= textureViewHeight) {
133                    bigEnough.add(option);
134                } else {
135                    notBigEnough.add(option);
136                }
137            }
138        }
139        //选择足够大的最小尺寸。如果没有足够大的尺寸，就在其中挑选最大的尺寸
140        if (bigEnough.size() > 0) {
141            return Collections.min(bigEnough, new Util.CompareSizesByArea());
```

```
142         } else if (notBigEnough.size() > 0) {
143             return Collections.max(notBigEnough, new Util.CompareSizesByArea());
144         } else {
145             return choices[0];
146         }
147     }
148     //比较两个 Size 大小的规则
149     static class CompareSizesByArea implements Comparator<Size> {
150         @Override
151         public int compare(Size lhs, Size rhs) {
152             return Long.signum((long) lhs.getWidth() * lhs.getHeight() - (long) rhs.getWidth() * rhs.getHeight());
153     }
154 }
```

第 107~112 行是 bitmap 实例转 byte[]实例的静态方法,用于隐式启动的回调数据传递。第 114~147 行是根据摄像头支持的尺寸、预览控件尺寸和最大预览尺寸限制,获取预览控件最佳尺寸的方法。第 149~154 行是用于比较 Size 实例大小运算规则的内置类,第 141 行和第 143 行是比较大小的运算规则。

3. 主界面的 Activity

```
/java/com/vt/c0805/MainActivity.java
72     //TextureView 的生命周期事件
73     private TextureView.SurfaceTextureListener mSurfaceTextureListener = new TextureView.SurfaceTextureListener() {
74         @Override
75         public void onSurfaceTextureAvailable(SurfaceTexture texture, int width, int height) {
76             Log.d(TAG, "TextureView.onSurfaceTextureAvailable()");
77             openCamera(width, height);
78         }
79         @Override
80         public void onSurfaceTextureSizeChanged(SurfaceTexture texture, int width, int height) { }
81         @Override
82         public void onSurfaceTextureUpdated(SurfaceTexture texture) { }
83         @Override
84         public boolean onSurfaceTextureDestroyed(SurfaceTexture texture) {
85             return true;
86         }
87     };
88     //摄像头设备状态的回调
89     private CameraDevice.StateCallback mCameraDeviceStateCallback = new CameraDevice.StateCallback() {
90         @Override
91         public void onOpened(@NonNull CameraDevice cameraDevice) {
92             Log.d(TAG, "CameraDevice.onOpened()");
93             mCameraDevice = cameraDevice;
94             startPreview();
95             mSemaphore.release();//释放 1 个信号许可
96         }
97         @Override
```

```java
98      public void onDisconnected(@NonNull CameraDevice cameraDevice) {
99          Log.d(TAG, "CameraDevice.onDisconnected()");
100         mSemaphore.release();//释放1个信号许可
101         cameraDevice.close();
102         mCameraDevice = null;
103     }
104     @Override
105     public void onError(@NonNull CameraDevice cameraDevice, int error) {
106         Log.d(TAG, "CameraDevice.onError()");
107         mSemaphore.release();//释放1个信号许可
108         cameraDevice.close();
109         mCameraDevice = null;
110     }
111 };
112 //ImageReader的监听器
113 private ImageReader.OnImageAvailableListener mOnImageAvailableListener = new ImageReader.OnImageAvailableListener() {
114     @Override
115     public void onImageAvailable(final ImageReader reader) {
116         Log.d(TAG, "ImageReader.onImageAvailable()");
117         //解除mImageReader的监听器
118         mImageReader.setOnImageAvailableListener(null, null);
119         //ImageReader读取到图像后使用新线程处理图像数据
120         ((Activity) mContext).runOnUiThread(new Runnable() {
121             @Override
122             public void run() {
123                 handlePhoto(reader);
124             }
125         });
126     }
127 };
128 @Override
129 protected void onCreate(Bundle savedInstanceState) {
130     super.onCreate(savedInstanceState);
131     setContentView(R.layout.activity_main);
132     //初始化
133     this.setTitle("C0805: 使用Camera2类拍摄照片");
134     mPreviewView = findViewById(R.id.texture_view);
135     mButton = findViewById(R.id.button);
136     mButton.setOnClickListener(this);
137 }
138 @Override
139 public void onResume() {
140     super.onResume();
141     startBackgroundThread();
142     //预览显示时打开摄像头，否则通过监听器监测打开摄像头
143     if (mPreviewView.isAvailable()) {
144         openCamera(mPreviewView.getWidth(), mPreviewView.getHeight());
145     } else {
146         mPreviewView.setSurfaceTextureListener(mSurfaceTextureListener);
```

```java
147         }
148     }
149     @Override
150     public void onPause() {
151         closeCamera();
152         stopBackgroundThread();
153         super.onPause();
154     }
155     //开启后台线程
156     private void startBackgroundThread() {
157         Log.d(TAG, "startBackgroundThread()");
158         mBackgroundThread = new HandlerThread("CameraBackground");
159         mBackgroundThread.start();
160         mBackgroundHandler = new Handler(mBackgroundThread.getLooper());
161     }
162     //停止后台线程
163     private void stopBackgroundThread() {
164         Log.d(TAG, "stopBackgroundThread()");
165         mBackgroundThread.quitSafely();
166         try {
167             mBackgroundThread.join();
168             mBackgroundThread = null;
169             mBackgroundHandler = null;
170         } catch (InterruptedException e) {
171             e.printStackTrace();
172         }
173     }
174     //单击事件
175     @Override
176     public void onClick(View view) {
177         takePhoto();
178     }
179     //打开摄像头
180     @SuppressLint("MissingPermission")
181     private void openCamera(int width, int height) {
182         Log.d(TAG, "openCamera()");
183         if (!Permissions.hasPermissionsGranted(mContext, PERMISSIONS)) {
184             Permissions.requestPermissions(mContext, PERMISSIONS);
185             return;
186         }
187         try {
188             //在2500ms内请求获取1个许可,否则抛出异常
189             if (!mSemaphore.tryAcquire(2500, TimeUnit.MILLISECONDS)) {
190                 throw new RuntimeException("打开摄像头超时");
191             }
192             mCameraManager = (CameraManager) getSystemService(Context.CAMERA_SERVICE);
193             //获取摄像头支持的属性特征
194             CameraCharacteristics characteristics = mCameraManager.getCameraCharacteristics(mCameraId);
195             //获取摄像头支持的可用流配置,包括每种格式、大小组合的最小帧持续时间和停顿持续时间
196             StreamConfigurationMap map = characteristics.get(CameraCharacteristics.SCALER_
```

```
                STREAM_CONFIGURATION_MAP);
197             //获取摄像头捕捉的最大尺寸读取照片
198             Size largest = Collections.max(Arrays.asList(map.getOutputSizes(ImageFormat.
        JPEG)), new Util.CompareSizesByArea());
199             mImageReader = ImageReader.newInstance(largest.getWidth(), largest.getHeight(),
        ImageFormat.JPEG, 2);
200             //选择最佳预览尺寸
201             mPreviewSize = Util.chooseOptimalSize(mContext, map.getOutputSizes(SurfaceTexture.class),
        width, height, MAX_PREVIEW_WIDTH, MAX_PREVIEW_HEIGHT, largest);
202             //设置预览尺寸
203             mPreviewView.setAspectRatio(mPreviewSize.getHeight(), mPreviewSize.getWidth());
204             //打开摄像头
205             mCameraManager.openCamera(mCameraId, mCameraDeviceStateCallback, mBackgroundHandler);
206         } catch (CameraAccessException e) {
207             e.printStackTrace();
208         } catch (InterruptedException e) {
209             throw new RuntimeException("打开摄像头被中断", e);
210         }
211     }
212     //创建摄像头预览会话
213     private void startPreview() {
214         Log.d(TAG, "startPreview()");
215         if (null == mCameraDevice || !mPreviewView.isAvailable() || null == mPreviewSize) { return; }
216         try {
217             closePreview();
218             //设置预览的缓冲区大小
219             SurfaceTexture texture = mPreviewView.getSurfaceTexture();
220             texture.setDefaultBufferSize(mPreviewSize.getWidth(), mPreviewSize.getHeight());
221             //设置预览输出的 Surface
222             Surface surface = new Surface(texture);
223             mCaptureRequestBuilder = mCameraDevice.createCaptureRequest(CameraDevice.
        TEMPLATE_PREVIEW);
224             mCaptureRequestBuilder.addTarget(surface);
225             //创建摄像头的捕获会话
226             mCameraDevice.createCaptureSession(Arrays.asList(surface, mImageReader.getSurface()),
        new CameraCaptureSession.StateCallback() {
227                 @Override
228                 public void onConfigured(CameraCaptureSession cameraCaptureSession) {
229                     mCaptureSession = cameraCaptureSession;
230                     updatePreview();
231                 }
232                 @Override
233                 public void onConfigureFailed(CameraCaptureSession cameraCaptureSession) {
234                     Toast.makeText(mContext, "摄像头配置失败", Toast.LENGTH_LONG).show();
235                 }
236             }, mBackgroundHandler);
237         } catch (CameraAccessException e) {
238             e.printStackTrace();
239         }
240     }
```

```java
241     //拍摄照片
242     private void takePhoto() {
243         Log.d(TAG, "takePhoto()");
244         if (null == mCameraDevice || !mPreviewView.isAvailable() || null == mPreviewSize) {
245             return;
246         }
247         try {
248             closePreview();
249             mCaptureRequestBuilder = mCameraDevice.createCaptureRequest(CameraDevice.TEMPLATE_STILL_CAPTURE);
250             //设置预览的缓冲区大小
251             SurfaceTexture texture = mPreviewView.getSurfaceTexture();
252             texture.setDefaultBufferSize(mPreviewSize.getWidth(), mPreviewSize.getHeight());
253             //设置预览的 Surface
254             List<Surface> surfaces = new ArrayList<>();
255             Surface previewSurface = new Surface(texture);
256             surfaces.add(previewSurface);
257             mCaptureRequestBuilder.addTarget(previewSurface);
258             //设置拍照的 Surface
259             Surface imageReaderSurface = mImageReader.getSurface();
260             surfaces.add(imageReaderSurface);
261             mCaptureRequestBuilder.addTarget(imageReaderSurface);
262             //创建捕捉会话
263             mCameraDevice.createCaptureSession(surfaces, new CameraCaptureSession.StateCallback() {
264                 @Override
265                 public void onConfigured(@NonNull CameraCaptureSession cameraCaptureSession) {
266                     Log.d(TAG, "CameraCaptureSession.onConfigured()");
267                     mCaptureSession = cameraCaptureSession;
268                     //添加 ImageReader 监听器处理拍摄照片
269                     mImageReader.setOnImageAvailableListener(mOnImageAvailableListener, mBackgroundHandler);
270                     mCaptureRequest = mCaptureRequestBuilder.build();
271                     try {//拍摄照片
272                         mCaptureSession.capture(mCaptureRequest, null, null);
273                     } catch (CameraAccessException e) {
274                         e.printStackTrace();
275                     }
276                 }
277                 @Override
278                 public void onConfigureFailed(CameraCaptureSession cameraCaptureSession) {
279                     Toast.makeText(mContext, "摄像头配置失败", Toast.LENGTH_LONG).show();
280                 }
281             }, mBackgroundHandler);
282         } catch (CameraAccessException e) {
283             e.printStackTrace();
284         }
285     }
286     //处理拍摄的照片
287     private void handlePhoto(ImageReader reader){
288         Log.d(TAG, "handlePhoto()");
```

```java
289        byte[] photoByte = Util.imageToBytes(reader.acquireNextImage());
290        mPhotoFile    =    Util.createFile(mContext.getExternalMediaDirs()[0].getAbsolutePath(),
    mContext.getResources().getString(R.string.app_name), "jpg");
291        //保存拍摄的照片
292        photoByte = Util.saveImage(photoByte, mPhotoFile, 90);
293        //将拍摄的照片显示在系统相册中
294        Util.showInAlbum(mContext, mPhotoFile.getAbsolutePath());
295        //恢复预览
296        startPreview();
297        //获取启动该 Activity 的 Intent
298        Intent intent = getIntent();
299        //判断是否通过外部 App 隐式启动
300        if (intent.getAction().equals("android.media.action.IMAGE_CAPTURE")
                ||intent.getAction().equals("android.media.action.STILL_IMAGE_CAMERA")) {
301            //判断是否指定了照片保存路径的 URI
302            Uri uri = intent.getParcelableExtra(MediaStore.EXTRA_OUTPUT);
303            if (uri != null) {
304                ContentResolver resolver = mContext.getContentResolver();
305                try {
306                    //向 URI 指定路径的文件写入照片数据
307                    ParcelFileDescriptor descriptor = resolver.openFileDescriptor(uri, "rw");
308                    FileOutputStream output = new FileOutputStream(descriptor.getFileDescriptor());
309                    output.write(photoByte);
310                    descriptor.close();
311                    output.close();
312                } catch (FileNotFoundException e) {
313                    e.printStackTrace();
314                } catch (IOException e) {
315                    e.printStackTrace();
316                }
317                setResult(RESULT_OK);
318                finish();
319            } else {
320                //设置照片压缩的参数
321                BitmapFactory.Options options = new BitmapFactory.Options();
322                options.inPreferredConfig = Bitmap.Config.RGB_565;
323                options.inSampleSize = 16;
324                //压缩照片
325                Bitmap bitmap = BitmapFactory.decodeFile(mPhotoFile.getAbsolutePath(), options);
326                //返回RESULT_OK，并包含一个 Intent 对象，Extra 中 key 为 data，value 是保存照片的 bitmap 对象
327                setResult(RESULT_OK, new Intent().putExtra("data", bitmap));
328                finish();
329            }
330        } else {
331            //启动预览照片的 Activity
332            intent = new Intent(MainActivity.this, PreviewActivity.class);
333            intent.putExtra("path", mPhotoFile.getAbsolutePath());
334            startActivity(intent);
335        }
336    }
```

```
337    //更新预览
338    private void updatePreview() {
339        try {
340            //设置自动对焦
341            mCaptureRequestBuilder.set(CaptureRequest.CONTROL_AF_MODE,    CaptureRequest.CONTROL_
       AF_MODE_CONTINUOUS_PICTURE);
342            mCaptureRequestBuilder.set(CaptureRequest.CONTROL_MODE,     CameraMetadata.CONTROL_
       MODE_AUTO);
343            //显示预览
344            mCaptureRequest = mCaptureRequestBuilder.build();
345            //捕捉图像会话设置重复请求
346            mCaptureSession.setRepeatingRequest(mCaptureRequest, null, mBackgroundHandler);
347        } catch (CameraAccessException e) {
348            e.printStackTrace();
349        }
350    }
351    //关闭预览会话
352    private void closePreview() {
353        if (mCaptureSession != null) {
354            mCaptureSession.close();
355            mCaptureSession = null;
356        }
357    }
358    //关闭摄像头
359    private void closeCamera() {
360        Log.d(TAG, "closeCamera()");
361        if (null != mCaptureSession) {
362            mCaptureSession.close();
363            mCaptureSession = null;
364        }
365        if (null != mCameraDevice) {
366            mCameraDevice.close();
367            mCameraDevice = null;
368        }
369        if (null != mImageReader) {
370            mImageReader.close();
371            mImageReader = null;
372        }
373        mSemaphore.release();//释放1个信号许可
374    }
```

第 73~87 行是 TextureView 控件的监听器,当 TextureView 控件的 SurfaceTexture 可用时调用 openCamera(int, int)方法。第 89~111 行是摄像头状态的回调,当摄像头打开后开始预览摄像头的画面,当摄像头断开连接或出现错误时关闭设备。第 113~127 行是 ImageReader 的图像可用监听器,当 ImageReader 的图像可用时,开启新线程调用 handlePhoto(reader)方法处理获取的图像数据。第 156~161 行创建后台线程,mBackgroundHandler 作为打开摄像头和图像可用监听器的线程,以避免主线程的阻塞。第

263~285 行创建摄像头的捕捉会话，当配置完成后捕捉图像，通过 mCaptureRequest 对象将捕捉的图像数据传递给 mImageReader。

扩展实例：Camera 类

Camera2 类的使用较为复杂，为了帮助读者理解及向下兼容的需要，在本书配套资源的"基础工程"文件夹中，提供了使用 Camera 类实现拍照的 C0803 工程，该工程与 C0805 工程实现的效果基本相同。

8.2.5 实例工程：使用 Camera2 类录制视频

本实例演示了使用 Camera2 类开发第三方视频录制 App 的方法，单击"录制"按钮后开始录制视频，此时按钮文字显示为"停止"，再次单击停止录制（如图 8-5 所示）。在其他 App 调用系统相机录制视频时，该 App 可以作为备选项供用户选择。

图 8-5 运行效果

1．打开基础工程

打开"基础工程"文件夹中的"C0806"工程，该工程已经包含 MainActivity、PreviewActivity、Permissions、PreviewTextureView、Util 类及布局和权限路径（/res/xml/file_paths.xml）的文件，在 AndroidManifest.xml 文件中还添加了录制和存储视频所需的权限及 FileProvider 权限。

2．Util 类

打开 Util 类，添加 chooseVideoSize()方法，用于选取录制视频的尺寸。

```
/java/com/vt/c0806/Util.java
148    //选取视频尺寸
149    public static Size chooseVideoSize(Size[] choices) {
150        for (Size size : choices) {
```

```
151              System.out.println(size.getWidth()+"*"+size.getHeight());
152              if (size.getWidth() == size.getHeight() * 4 / 3 && size.getWidth() <= 1080) {
153                  return size;
154              }
155          }
156          return choices[choices.length - 1];
157      }
```

第 152 行判断视频尺寸是不是 4:3 比例画幅,并且宽度不大于 1080 像素,使用虚拟设备运行时,如果比例画幅设置为 16:9,可能会出现卡顿现象,使用物理设备运行时推荐使用 16:9 比例画幅。

3. 主界面的 Activity

```
/java/com/vt/c0806/MainActivity.java
66   //TextureView 的生命周期事件
67   private TextureView.SurfaceTextureListener mSurfaceTextureListener = new
     TextureView.SurfaceTextureListener() {
68       @Override
69       public void onSurfaceTextureAvailable(SurfaceTexture surfaceTexture, int width, int height) {
70           Log.d(TAG, "TextureView.onSurfaceTextureAvailable()");
71           openCamera(width, height);
72       }
73       @Override
74       public void onSurfaceTextureSizeChanged(SurfaceTexture surfaceTexture, int width, int height) { }
75       @Override
76       public void onSurfaceTextureUpdated(SurfaceTexture surfaceTexture) { }
77       @Override
78       public boolean onSurfaceTextureDestroyed(SurfaceTexture surfaceTexture) {
79           return true;
80       }
81   };
82   //摄像头设备状态的回调
83   private CameraDevice.StateCallback mCameraDeviceStateCallback = new CameraDevice.
     StateCallback() {
84       @Override
85       public void onOpened(@NonNull CameraDevice cameraDevice) {
86           Log.d(TAG, "CameraDevice.onOpened()");
87           mCameraDevice = cameraDevice;
88           startPreview();
89           mSemaphore.release();//释放 1 个信号许可
90       }
91       @Override
92       public void onDisconnected(@NonNull CameraDevice cameraDevice) {
93           Log.d(TAG, "CameraDevice.onDisconnected()");
94           mSemaphore.release();//释放 1 个信号许可
95           cameraDevice.close();
96           mCameraDevice = null;
97       }
98       @Override
99       public void onError(@NonNull CameraDevice cameraDevice, int error) {
```

```java
100                Log.d(TAG, "CameraDevice.onError()");
101                mSemaphore.release();//释放1个信号许可
102                cameraDevice.close();
103                mCameraDevice = null;
104            }
105        };
106        @Override
107        protected void onCreate(Bundle savedInstanceState) {
108            super.onCreate(savedInstanceState);
109            setContentView(R.layout.activity_main);
110            //初始化
111            this.setTitle("C0806: 使用Camera2类录制视频");
112            mPreviewView = findViewById(R.id.texture_view);
113            mButton = findViewById(R.id.button);
114            mButton.setOnClickListener(this);
115        }
116        @Override
117        public void onResume() {
118            super.onResume();
119            startBackgroundThread();
120            //预览显示时打开摄像头，否则通过监听器监测打开摄像头
121            if (mPreviewView.isAvailable()) {
122                openCamera(mPreviewView.getWidth(), mPreviewView.getHeight());
123            } else {
124                mPreviewView.setSurfaceTextureListener(mSurfaceTextureListener);
125            }
126        }
127        @Override
128        public void onPause() {
129            closeCamera();
130            stopBackgroundThread();
131            super.onPause();
132        }
133        //开启后台线程
134        private void startBackgroundThread() {
135            Log.d(TAG, "startBackgroundThread()");
136            mBackgroundThread = new HandlerThread("CameraBackground");
137            mBackgroundThread.start();
138            mBackgroundHandler = new Handler(mBackgroundThread.getLooper());
139        }
140        //停止后台线程
141        private void stopBackgroundThread() {
142            Log.d(TAG, "stopBackgroundThread()");
143            mBackgroundThread.quitSafely();
144            try {
145                mBackgroundThread.join();
146                mBackgroundThread = null;
147                mBackgroundHandler = null;
148            } catch (InterruptedException e) {
149                e.printStackTrace();
```

```
150        }
151    }
152    @Override
153    public void onClick(View v) {
154        if (mIsRecordingVideo) {
155            stopRecordingVideo(false);
156        } else {
157            startRecordingVideo();
158        }
159    }
160    //打开摄像头
161    @SuppressWarnings("MissingPermission")
162    private void openCamera(int width, int height) {
163        Log.d(TAG, "openCamera()");
164        if (!Permissions.hasPermissionsGranted(mContext, PERMISSIONS)) {
165            Permissions.requestPermissions(mContext, PERMISSIONS);
166            return;
167        }
168        try {
169            //在2500ms内请求获取1个许可,否则抛出异常
170            if (!mSemaphore.tryAcquire(2500, TimeUnit.MILLISECONDS)) {
171                throw new RuntimeException("打开摄像头超时");
172            }
173            mCameraManager = (CameraManager) mContext.getSystemService(Context.CAMERA_SERVICE);
174            //获取后置摄像头
175            String cameraId = mCameraManager.getCameraIdList()[0];
176            //获取预览和录制视频的尺寸
177            CameraCharacteristics characteristics = mCameraManager.getCameraCharacteristics(cameraId);
178            StreamConfigurationMap map = characteristics.get(CameraCharacteristics.SCALER_STREAM_CONFIGURATION_MAP);
179            mVideoSize = Util.chooseVideoSize(map.getOutputSizes(MediaRecorder.class));
180            mPreviewSize = Util.chooseOptimalSize(mContext, map.getOutputSizes(SurfaceTexture.class), width, height, MAX_PREVIEW_WIDTH, MAX_PREVIEW_HEIGHT, mVideoSize);
181            mPreviewView.setAspectRatio(mPreviewSize.getHeight(), mPreviewSize.getWidth());
182            mMediaRecorder = new MediaRecorder();
183            mCameraManager.openCamera(cameraId, mCameraDeviceStateCallback, mBackgroundHandler);
184        } catch (CameraAccessException e) {
185            Toast.makeText(mContext, "摄像头不可用", Toast.LENGTH_LONG).show();
186        } catch (InterruptedException e) {
187            throw new RuntimeException("打开摄像头被中断", e);
188        }
189    }
190    //设置录制参数
191    private void setUpMediaRecorder() throws IOException {
192        Log.d(TAG, "setUpMediaRecorder()");
193        mVideoFile = Util.createFile(mContext.getExternalMediaDirs()[0].getAbsolutePath(), mContext.getResources().getString(R.string.app_name), "mp4");
194        mMediaRecorder.setVideoSource(MediaRecorder.VideoSource.SURFACE);
195        mMediaRecorder.setAudioSource(MediaRecorder.AudioSource.MIC);
196        mMediaRecorder.setOutputFormat(MediaRecorder.OutputFormat.MPEG_4);
```

```java
197     mMediaRecorder.setOutputFile(mVideoFile.getAbsolutePath());
198     mMediaRecorder.setVideoEncoder(MediaRecorder.VideoEncoder.H264);
199     mMediaRecorder.setAudioEncoder(MediaRecorder.AudioEncoder.AAC);
200     mMediaRecorder.setVideoEncodingBitRate(8 * mVideoSize.getWidth() * mVideoSize.getHeight());
201     mMediaRecorder.setVideoFrameRate(30);
202     mMediaRecorder.setVideoSize(mVideoSize.getWidth(), mVideoSize.getHeight());
203     mMediaRecorder.setOrientationHint(90);//设置视频的角度
204     mMediaRecorder.setOnErrorListener(new MediaRecorder.OnErrorListener() {
205         @Override
206         public void onError(MediaRecorder mr, int what, int extra) {
207             stopRecordingVideo(true);
208         }
209     });
210     mMediaRecorder.prepare();
211 }
212 //开始预览
213 private void startPreview() {
214     Log.d(TAG, "startPreview()");
215     if (null == mCameraDevice || !mPreviewView.isAvailable() || null == mPreviewSize) {
216         return;
217     }
218     try {
219         closePreview();
220         //设置预览的缓冲区大小
221         SurfaceTexture texture = mPreviewView.getSurfaceTexture();
222         texture.setDefaultBufferSize(mPreviewSize.getWidth(), mPreviewSize.getHeight());
223         //设置预览输出的 Surface
224         Surface previewSurface = new Surface(texture);
225         mCaptureRequestBuilder = mCameraDevice.createCaptureRequest(CameraDevice.TEMPLATE_PREVIEW);
226         mCaptureRequestBuilder.addTarget(previewSurface);
227         //创建摄像头的捕获会话
228         mCameraDevice.createCaptureSession(Collections.singletonList(previewSurface),
    new CameraCaptureSession.StateCallback() {
229             @Override
230             public void onConfigured(@NonNull CameraCaptureSession session) {
231                 mCaptureSession = session;
232                 updatePreview();
233             }
234             @Override
235             public void onConfigureFailed(@NonNull CameraCaptureSession session) {
236                 Toast.makeText(mContext, "摄像头配置失败", Toast.LENGTH_LONG).show();
237             }
238         }, mBackgroundHandler);
239     } catch (CameraAccessException e) {
240         e.printStackTrace();
241     }
242 }
243 //开始录制视频
244 private void startRecordingVideo() {
```

```java
245         Log.d(TAG, "startRecordingVideo()");
246         if (null == mCameraDevice || !mPreviewView.isAvailable() || null == mPreviewSize) {
247             return;
248         }
249         try {
250             closePreview();
251             setUpMediaRecorder();
252             mCaptureRequestBuilder = mCameraDevice.createCaptureRequest(CameraDevice.TEMPLATE_RECORD);
253             //设置预览的缓冲区大小
254             SurfaceTexture texture = mPreviewView.getSurfaceTexture();
255             texture.setDefaultBufferSize(mPreviewSize.getWidth(), mPreviewSize.getHeight());
256             //设置预览的 Surface
257             List<Surface> surfaces = new ArrayList<>();
258             Surface previewSurface = new Surface(texture);
259             surfaces.add(previewSurface);
260             mCaptureRequestBuilder.addTarget(previewSurface);
261             //设置录制视频的 Surface
262             Surface recorderSurface = mMediaRecorder.getSurface();
263             surfaces.add(recorderSurface);
264             mCaptureRequestBuilder.addTarget(recorderSurface);
265             //创建捕捉会话
266             mCameraDevice.createCaptureSession(surfaces, new CameraCaptureSession.StateCallback() {
267                 @Override
268                 public void onConfigured(@NonNull CameraCaptureSession session) {
269                     mCaptureSession = session;
270                     updatePreview();
271                     ((Activity) mContext).runOnUiThread(new Runnable() {
272                         @Override
273                         public void run() {
274                             mButton.setText("停止");
275                             mIsRecordingVideo = true;
276                             mMediaRecorder.start();
277                         }
278                     });
279                 }
280                 @Override
281                 public void onConfigureFailed(CameraCaptureSession cameraCaptureSession) {
282                     Toast.makeText(mContext, "摄像头配置失败", Toast.LENGTH_LONG).show();
283                 }
284             }, mBackgroundHandler);
285         } catch (CameraAccessException | IOException e) {
286             e.printStackTrace();
287         }
288     }
289     //停止录制视频
290     private void stopRecordingVideo(boolean error) {
291         mMediaRecorder.stop();
292         mMediaRecorder.reset();
293         mIsRecordingVideo = false;
294         mButton.setText("录制");
```

```java
295        startPreview();
296        if (error) {//如果因为错误停止，则直接返回不进行后续处理
297            return;
298        }
299        Util.showInAlbum(mContext, mVideoFile.getAbsolutePath());
300        Uri videoUri = FileProvider.getUriForFile(this, "com.vt.c0806.fileprovider", mVideoFile);
301        Intent intent = getIntent();
302        //判断是否被其他 App 隐式启动
303        if (intent.getAction().equals("android.media.action.VIDEO_CAPTURE")) {
304            intent = new Intent();
305            intent.setDataAndType(videoUri, "video/mp4");//设置返回的数据及其类型
306            intent.setFlags(Intent.FLAG_GRANT_READ_URI_PERMISSION);//授予临时读取权限
307            setResult(RESULT_OK, intent);
308            //关闭返回其他 App
309            finish();
310        } else {
311            //启动播放录制视频的 Activity
312            intent = new Intent(MainActivity.this, PreviewActivity.class);
313            intent.setDataAndType(videoUri, "video/mp4");
314            startActivity(intent);
315        }
316    }
317    //更新预览
318    private void updatePreview() {
319        try {
320            //设置自动对焦
321            mCaptureRequestBuilder.set(CaptureRequest.CONTROL_MODE, CameraMetadata.CONTROL_MODE_AUTO);
322            //显示预览
323            mCaptureRequest = mCaptureRequestBuilder.build();
324            HandlerThread thread = new HandlerThread("CameraPreview");
325            thread.start();
326            //捕捉图像会话设置重复请求
327            mCaptureSession.setRepeatingRequest(mCaptureRequest, null, mBackgroundHandler);
328        } catch (CameraAccessException e) {
329            e.printStackTrace();
330        }
331    }
332    //关闭预览会话
333    private void closePreview() {
334        if (mCaptureSession != null) {
335            mCaptureSession.close();
336            mCaptureSession = null;
337        }
338    }
339    //关闭摄像头
340    private void closeCamera() {
341        closePreview();
342        if (null != mCaptureSession) {
343            mCaptureSession.close();
344            mCaptureSession = null;
```

```
345         }
346         if (null != mCameraDevice) {
347             mCameraDevice.close();
348             mCameraDevice = null;
349         }
350         if (null != mMediaRecorder) {
351             mMediaRecorder.release();
352             mMediaRecorder = null;
353         }
354         mSemaphore.release();//释放 1 个信号许可
355     }
```

第 67～81 行是用于 TextureView 控件的监听器，当 TextureView 控件的 SurfaceTexture 可用时调用 openCamera(int, int)方法。第 83～105 行是摄像头状态的回调，当摄像头打开后开始预览摄像头的画面，当摄像头断开连接或出现错误时关闭设备。第 134～139 行创建后台线程，mBackgroundHandler 作为打开摄像头、预览捕捉会话和录制捕捉会话的线程，以避免主线程的阻塞。第 228～241 行创建摄像头的预览视频捕捉会话，当配置完成后，使用 updatePreview()方法更新预览图像，实现视频的预览。第 266～287 行创建摄像头的录制视频捕捉会话，当配置完成后，调用 mMediaRecorder.start()方法录制视频。

扩展实例：Camera 类录制视频

Camera 类的使用较为复杂，为了帮助读者理解及向下兼容的需要，在配套资源的"基础工程"文件夹中提供了使用 Camera 类实现录制视频的 C0804 工程，该工程与 C0806 工程实现效果基本相同。

8.3 录 制 音 频

麦克风采集到的音频可以通过 MediaRecorder 类或 AudioRecord 类进行录制。MediaRecorder 类录制音频的方法和录制视频的方法是类似的，只需要在录制参数中删除视频相关的内容即可。而 AudioRecord 类提供的方法更加丰富，能够直接获取采集的音频流数据。音频流数据是原始的 PCM 格式音频数据，保存到文件后，需要添加头信息才能转为可以播放的 WAV 格式文件。MP3 等压缩格式的音频文件需要进行转码，转为 MP3 格式通常使用第三方类库——Lame 库。

播放音频可以使用 MediaPlayer 类、AudioTrack 类、SoundPool 类或 Ringtone 类，它们的使用场景有所不同。MediaPlayer 类播放音频和播放视频的方法是一样的，支持添加 MediaController 类的控制器；AudioTrack 类用于播放 PCM 音乐流，支持流模式，播放其他压缩格式需要先进行解码；SoundPool 类用于多个短音频文件交错密集播放，多用于游戏、乐器类 App。Ringtone 类用于铃声、通知声的播放，可以通过 RingtoneManager 类获取系统铃声列表。

8.3.1 AudioRecord 类

AudioRecord 类（android.media.AudioRecord）用于读取麦克风采集的音频流数据（如

表 8-15 所示）。创建后，AudioRecord 对象将初始化其关联的音频缓冲区，它将用新的音频数据填充。在构造期间指定的此缓冲区的大小确定 AudioRecord 在尚未读取的"超速运行"数据之前可以记录多长时间。注意，应从音频硬件读取数据，其大小应小于总记录缓冲区的大小。

表 8-15 AudioRecord 类的常用方法

类型和修饰符	方 法
	AudioRecord(int audioSource, int sampleRateInHz, int channelConfig, int audioFormat, int bufferSizeInBytes) 构造方法。audioSource 参数使用以下 MediaRecorder.AudioSource 类定义的常量表示不同的音源模式。 ● DEFAULT：默认模式 ● MIC：麦克风模式 ● VOICE_UPLINK：电话上行音频模式 ● VOICE_DOWNLINK：电话下行音频模式 ● VOICE_CALL：电话模式 ● CAMCORDER：麦克风方向模式，根据开启的摄像头方向选择 ● VOICE_RECOGNITION：识别模式，先进行声音识别，然后录制 ● VOICE_COMMUNICATION：交流模式，开启回声消除和自动增益 ● REMOTE_SUBMIX：远程混合模式，录制系统内置音频
int	getAudioSource() 获取音源
AudioFormat	getFormat() 获取音频录制的格式
static int	getMinBufferSize(int sampleRateInHz, int channelConfig, int audioFormat) 获取最小缓冲区大小
int	getSampleRate() 获取音频采样率
int	read(byte[] audioData, int offsetInBytes, int sizeInBytes) 从音频输入设备读取音频流数据保存到 byte 数组中
void	release() 释放 AudioRecord 实例的占用资源
void	startRecording() 开始录制音频
void	stop() 停止录制音频

8.3.2 AudioTrack 类

AudioTrack 类（android.media.AudioTrack）用于管理和播放单个音频资源（如表 8-16 所示），支持静态和流式两种方式播放音频。

表 8-16 AudioTrack 类的常用方法

类型和修饰符	方 法
	AudioTrack(AudioAttributes attributes, AudioFormat format, int bufferSizeInBytes, int mode, int sessionId) 构造方法
static float	getMaxVolume() 获取最大音量
static int	getMinBufferSize(int sampleRateInHz, int channelConfig, int audioFormat) 获取所需的最小缓冲区

续表

类型和修饰符	方　　法
static float	getMinVolume() 获取最小音量
int	getPlaybackRate() 获取回放采样频率
int	getSampleRate() 获取音频采样频率
void	pause() 暂停播放音频数据
void	play() 播放音频数据
void	release() 释放资源
int	setPlaybackRate(int sampleRateInHz) 设置回放的采样速率
int	setVolume(float gain) 设置音量
void	stop() 停止播放音频
int	write(byte[] audioData, int offsetInBytes, int sizeInBytes) 写入播放的音频流数据

8.3.3 实例工程：AudioRecord 录制音频

本实例演示了使用不同方式通过 AudioRecord 录制音频，然后使用 AudioTrack 播放音频（如图 8-6 所示）。单击"开始录音"按钮后会显示已经录制的时间，底部显示录制音频文件保存的文件路径。该按钮文字变成"暂停"，单击后暂停录音，按钮文字变成"继续"，再次单击后继续录音。单击"结束录音"按钮停止录音，单击"播放录音"按钮播放录制的音频。由于虚拟设备无法使用麦克风，因此需要使用物理设备才能录制和播放音频。

图 8-6　运行效果

1. 打开基础工程

打开"基础工程"文件夹中的"C0807"工程,该工程已经包含 MainActivity、Permissions、Util 类及布局的文件,在 AndroidManifest.xml 文件中还添加了录音和存储文件所需的权限。Util 类中已经添加了两个可重用的静态方法——pcmToWave(String,String,long,int)方法和 writeWaveFileHeader(FileOutputStream, long, long, long, int, long)方法,用于转为 WAV 格式文件。

2. 主界面的 Activity

```
/java/com/vt/c0807/MainActivity.java
138    //单选按钮选项改变监听事件
139    @Override
140    public void onCheckedChanged(RadioGroup group, int checkedId) {
141        switch (checkedId) {
142            case R.id.default_radio_button:
143                mAudioSource = MediaRecorder.AudioSource.DEFAULT;
144                break;
145            case R.id.mic_radio_button:
146                mAudioSource = MediaRecorder.AudioSource.MIC;
147                break;
148            case R.id.call_radio_button:
149                mAudioSource = MediaRecorder.AudioSource.VOICE_CALL;
150                break;
151            case R.id.communication_radio_button:
152                mAudioSource = MediaRecorder.AudioSource.VOICE_COMMUNICATION;
153                break;
154        }
155    }
156    //开始录音
157    void startRecord() {
158        mIsStartRecording = true;
159        mIsPauseRecording = false;
160        mPlayButton.setEnabled(false);
161        mStartButton.setText("暂停");
162        mPCMFile = Util.createFile(this.getExternalFilesDir("audio").getAbsolutePath(), "", "pcm");
163        mWAVFile = Util.createFile(this.getExternalFilesDir("audio").getAbsolutePath(), "", "wav");
164        mLogTextView.setText(mPCMFile.getAbsolutePath());
165        mBufferSize = AudioRecord.getMinBufferSize(mInSampleRate, mInChannelConfig, mInAudioFormat);
166        mAudioData = new byte[mBufferSize];
167        mAudioRecord = new AudioRecord(mAudioSource, mInSampleRate, mInChannelConfig, mInAudioFormat, mBufferSize);
168        //开始录音(此后能够读取音频数据)
169        mAudioRecord.startRecording();
170        //开启计时线程
171        mTimerThread = new Thread(this);
172        mTimerThread.start();
173        //录制音频的线程
174        mExecutor.execute(new Runnable() {
175            @Override
176            public void run() {
```

```
177             try {
178                 FileOutputStream outputStream = new FileOutputStream(mPCMFile.getAbsoluteFile());
179                 //开始录制时循环
180                 while (mIsStartRecording) {
181                     //暂停时循环
182                     while (mIsStartRecording&&mIsPauseRecording) {
183                         try {
184                             Thread.sleep(100);
185                         } catch (InterruptedException e) {
186                             e.printStackTrace();
187                         }
188                     }
189                     mAudioRecord.read(mAudioData, 0, mAudioData.length);
190                     outputStream.write(mAudioData);
191                 }
192                 outputStream.close();
193                 //将 PCM 文件转为 WAV 文件
194                 Util.pcmToWave(mPCMFile.getAbsolutePath(), mWAVFile.getAbsolutePath(),
    mInSampleRate, mBufferSize);
195             } catch (FileNotFoundException e) {
196                 e.printStackTrace();
197             } catch (IOException e) {
198                 e.printStackTrace();
199             }
200         }
201     });
202 }
203 //暂停录音
204 void pauseRecord() {
205     mAudioRecord.stop();
206     mIsPauseRecording = true;
207     mStartButton.setText("继续");
208 }
209 //继续录音
210 void continueRecord() {
211     mAudioRecord.startRecording();
212     mIsPauseRecording = false;
213     mStartButton.setText("暂停");
214 }
215 //停止录音
216 void stopRecord() {
217     mAudioRecord.stop();
218     mIsStartRecording = false;
219     mIsPauseRecording = true;
220     mStartButton.setText("开始录音");
221     Toast.makeText(this, "录音已结束", Toast.LENGTH_SHORT).show();
222     mPlayButton.setEnabled(true);
223 }
224 //创建音轨
225 public void createAudioTrack() throws IllegalStateException {
```

```java
226     int mBufferSizeInBytes = AudioTrack.getMinBufferSize(mOutSampleRate, mOutChannelConfig,
    mOutAudioFormat);
227     if (mBufferSizeInBytes <= 0) {
228         throw new IllegalStateException("最小缓冲区尺寸: " + mBufferSizeInBytes);
229     }
230     //判断当前版本是否大于等于 Android M (API level 23)
231     if (Build.VERSION.SDK_INT >= Build.VERSION_CODES.M) {
232         mAudioTrack = new AudioTrack.Builder()
233                 .setAudioAttributes(new AudioAttributes.Builder()
234                         .setUsage(AudioAttributes.USAGE_MEDIA)
235                         .setContentType(AudioAttributes.CONTENT_TYPE_MUSIC)
236                         .setLegacyStreamType(AudioManager.STREAM_MUSIC)
237                         .build())
238                 .setAudioFormat(new AudioFormat.Builder()
239                         .setEncoding(mInAudioFormat)
240                         .setSampleRate(mInSampleRate)
241                         .setChannelMask(mInChannelConfig)
242                         .build())
243                 .setTransferMode(AudioTrack.MODE_STREAM)
244                 .setBufferSizeInBytes(mBufferSizeInBytes)
245                 .build();
246     } else {
247         mAudioTrack = new AudioTrack(AudioManager.STREAM_MUSIC, mOutSampleRate, mOutChannelConfig,
248             mOutAudioFormat, mBufferSizeInBytes, AudioTrack.MODE_STREAM);
249     }
250     mAudioTrack.setVolume(1.0f);
251 }
252 //播放音频
253 public void play() throws IllegalStateException {
254     mExecutor.execute(new Runnable() {
255         @Override
256         public void run() {
257             try {
258                 playAudioData(mWAVFile);
259             } catch (IOException e) {
260                 mMainHandler.post(new Runnable() {
261                     @Override
262                     public void run() {
263                         Toast.makeText(mContext, "录音播放出现错误", Toast.LENGTH_SHORT).show();
264                     }
265                 });
266             }
267         }
268     });
269 }
270 //播放音频数据
271 private void playAudioData(File audioFile) throws IOException {
272     FileInputStream fis = new FileInputStream(audioFile);
273     DataInputStream dis = new DataInputStream(new BufferedInputStream(fis));
274     byte[] bytes = new byte[mInSampleRate];
```

```
275     int len;
276     mAudioTrack.play();//播放音轨（此时不会播放出声音）
277     while ((len = dis.read(bytes)) != -1) {
278         mAudioTrack.write(bytes, 0, len);//写入音轨播放音频
279     }
280     mMainHandler.post(new Runnable() {
281         @Override
282         public void run() {
283             Toast.makeText(mContext, "录音播放完毕", Toast.LENGTH_SHORT).show();
284         }
285     });
286     if (dis != null) {
287         dis.close();//关闭释放资源
288     }
289 }
```

第 140～155 行是单选按钮选项改变监听事件，选择相应的选项后设置不同的音源类型。第 174～201 行通过 mExecutor 对象开启新线程循环读取音频流，直至停止录音，当暂停时执行内部循环暂停 100ms。第 225～251 行根据当前设备版本的不同使用两种方法实例化 mAudioTrack 对象创建音轨，并设置最大音量。第 271～289 行播放录制的音频文件，通过循环读取的方式播放音频数据。

8.3.4 实例工程：MediaRecorder 录制音频

本实例演示了使用不同方式通过 MediaRecorder 录制 AMR 格式音频，然后使用 MediaPlayer 播放音频（如图 8-7 所示）。单击"开始录音"按钮后会显示已经录制的时间，底部显示录制音频文件保存的文件路径。该按钮文字变成"暂停"，单击后暂停录音，按钮文字变成"继续"，再次单击后继续录音。单击"结束录音"按钮停止录音，单击"播放录音"按钮播放录制的音频。由于虚拟设备无法使用麦克风，因此需要使用物理设备才能录制和播放音频。

图 8-7　运行效果

提示：AMR 格式

AMR（Adaptive Multi-Rate）格式是欧洲通信标准化委员会提出的音频标准格式，其压缩比比 MP3 格式的大。

1. 打开基础工程

打开"基础工程"文件夹中的"C0808"工程，该工程已经包含 MainActivity、Permissions、Util 类及布局的文件，在 AndroidManifest.xml 文件中已经添加了录音和存储文件所需的权限。

2. 主界面的 Activity

```
/java/com/vt/c0808/MainActivity.java
113  //开始录音
114  private void startRecord(){
115      try {
116          if (!mIsStartRecording) {
117              mAudioFile = Util.createFile(this.getExternalFilesDir("audio").getAbsolutePath(),
     "", "amr");
118              mLogTextView.setText(mAudioFile.getAbsolutePath());
119              mMediaRecorder = new MediaRecorder();
120              mMediaRecorder.setAudioSource(mAudioSource);//音频输入源
121              mMediaRecorder.setOutputFormat(MediaRecorder.OutputFormat.AMR_WB);//设置输出格式
122              mMediaRecorder.setAudioEncoder(MediaRecorder.AudioEncoder.AMR_WB);//设置编码格式
123              mMediaRecorder.setOutputFile(mAudioFile.getAbsolutePath());//设置输出文件
124              mMediaRecorder.prepare();//录音准备
125              mIsStartRecording = true;
126              mIsPauseRecording = false;
127          }
128          if (mIsPauseRecording) {
129              mMediaRecorder.resume();//继续录音
130              mIsPauseRecording = false;
131          } else {
132              mMediaRecorder.start();//开始录音
133          }
134          mStartButton.setText("暂停");
135          //开启计时线程
136          mTimerThread = new Thread(this);
137          mTimerThread.start();
138      } catch (IOException e) {
139          e.printStackTrace();
140      }
141  }
142  //暂停录音
143  private void pauseRecord(){
144      if(mMediaRecorder != null){
145          mMediaRecorder.pause();//暂停录音
146          mIsPauseRecording = true;
147          mStartButton.setText("继续");
148      }
149  }
```

```
150    //停止录音
151    private void stopRecord(){
152        if(mMediaRecorder != null){
153            mMediaRecorder.stop();//停止录音
154            mMediaRecorder.release();
155            mMediaRecorder = null;
156            mIsStartRecording = false;
157            mIsPauseRecording = true;
158            mStartButton.setText("开始录音");
159            Toast.makeText(this, "录音结束", Toast.LENGTH_SHORT).show();
160        }
161    }
162    //播放音频
163    private void play(){
164        try {
165            Uri uri = Uri.fromFile(mAudioFile);
166            MediaPlayer mediaPlayer = new MediaPlayer();
167            mediaPlayer.setAudioStreamType(AudioManager.STREAM_MUSIC);
168            mediaPlayer.setDataSource(getApplicationContext(), uri);
169            mediaPlayer.prepare();
170            mediaPlayer.start();
171        } catch (IOException e) {
172            e.printStackTrace();
173        }
174    }
```

第 116～137 行在开始录音前配置相关的参数，使用 mMediaRecorder.start()方法开始录音或暂停后使用 mMediaRecorder.resume()方法继续录音，然后通过 mTimerThread 对象开启新线程计时。第 145 行使用 mMediaRecorder.pause()方法暂停录音。第 153 行使用 mMediaRecorder.stop()方法停止录音，停止录音后会自动保存设置的输出文件。第 165～170 行通过 mediaPlayer 对象播放指定的 URI 路径的录音文件。

8.4 传 感 器

传感器是用于感知外部环境数据的设备。在广义范畴上，摄像头、麦克风和 GPS 都属于传感器。Android 系统所指的传感器是狭义范畴的，是指除摄像头、麦克风和 GPS 外的能够感知外部环境数据的设备，这些传感器都能使用 SensorManager 类进行管理。

8.4.1 传感器概述

Android 设备内置的传感器用于测量运动、屏幕方向和各种环境条件，能够提供高度精确的原始数据用来监测 Android 设备的位置、运动或周围环境的变化。

SensorManager 类（android.hardware.SensorManager）用于获取和访问传感器，注册和注销传感器事件监听器，以及获取屏幕方向的数据（如表 8-17 所示）；还提供了几个传感器常量，用于报告传感器精确度，设置数据采集频率和校准传感器。

表 8-17 SensorManager 类的常用方法

类型和修饰符	方法
static float	getAltitude(float p0, float p) 获取海平面大气压力 p0 和指定大气压力 p 计算以 m 为单位的海拔高度
static void	getAngleChange(float[] angleChange, float[] R, float[] prevR) 计算两个旋转矩阵之间的角度变化，使用 angleChange 参数进行存储
Sensor	getDefaultSensor(int type) 获取指定类型的默认传感器
Sensor	getDefaultSensor(int type, boolean wakeUp) 获取具有指定类型和唤醒属性的传感器
static float[]	getOrientation(float[] R, float[] values) 根据旋转矩阵计算设备的方向，使用 values 参数进行存储
static boolean	getRotationMatrix(float[] R, float[] I, float[] gravity, float[] geomagnetic) 通过重力矩阵 gravity 和地磁矩阵 geomagnetic 从设备坐标系转换为世界坐标系的倾角矩阵 I 和旋转矩阵 R
static void	getRotationMatrixFromVector(float[] R, float[] rotationVector) 将旋转矢量转换为旋转矩阵，使用 R 参数进行存储
void	registerDynamicSensorCallback(SensorManager.DynamicSensorCallback callback) 注册动态传感器的回调
boolean	registerListener(SensorEventListener listener, Sensor sensor, int samplingPeriodUs) 注册指定采样频率的传感器监听器
void	unregisterDynamicSensorCallback(SensorManager.DynamicSensorCallback callback) 注销动态传感器的回调
void	unregisterListener(SensorEventListener listener) 注销传感器的监听器

Sensor 类（android.hardware.Sensor）用于获取和判断传感器的特性（如表 8-18 所示），需要通过 SensorManager.getDefaultSensor(int)方法或 SensorManager.getDefaultSensor(int, boolean)方法获取指定类型的传感器实例。传感器的类型使用 Sensor 类的常量表示，有些是指硬件传感器，有些是指通过单个或多个硬件传感器虚拟的软件传感器。

表 8-18 Sensor 类的常用方法

类型和修饰符	方法
int	getMaxDelay() 获取两个传感器事件之间的最大延迟，仅针对连续变化传感器
int	getMinDelay() 获取传感器测量数据更改时返回数值的最小延迟时间（μs）
String	getStringType() 获取传感器类型的字符串
int	getType() 获取传感器的类型
int	getVersion() 获取传感器的版本号

SensorEvent 类（android.hardware.SensorEvent）提供有关传感器事件的信息，包括发生事件的传感器数据、传感器类型、数据精度和时间戳（如表 8-19 所示）。

表 8-19　SensorEvent 类的属性

类型和修饰符	属　　性
public int	accuracy 事件的数据精度。数据精度的数值可使用以下常量表示。 ● SensorManager.SENSOR_STATUS_ACCURACY_HIGH：高精度值 ● SensorManager.SENSOR_STATUS_ACCURACY_LOW：低精度值 ● SensorManager.SENSOR_STATUS_ACCURACY_MEDIUM：中等精度值 ● SensorManager.SENSOR_STATUS_ACCURACY_UNRELIABLE：精度值不可靠
public Sensor	sensor 事件的传感器类型
public long	timestamp 事件的时间戳
public final float[]	values 事件的数据

SensorEventListener 接口（android.hardware.SensorEventListener）用于监听传感器获取的数值或传感器精确度的变化（如表 8-20 所示）。

表 8-20　SensorEventListener 接口的方法

类型和修饰符	方　　法
void	onAccuracyChanged(Sensor sensor, int accuracy) 当传感器的精度改变时，调用此方法
void	onSensorChanged(SensorEvent event) 当传感器值改变时，调用此方法

8.4.2　运动类传感器

运动类传感器用于测量设备的运动（如倾斜、晃动、旋转或摆动），包含加速度传感器、重力传感器、陀螺仪传感器和旋转矢量传感器（如表 8-21 所示）。

表 8-21　Sensor 类的运动类传感器

传感器类型	传感器事件数据	说　　明
TYPE_ACCELEROMETER 加速度传感器（单位：m/s^2）	SensorEvent.values[0]	沿 x 轴的加速力（包括重力）
	SensorEvent.values[1]	沿 y 轴的加速力（包括重力）
	SensorEvent.values[2]	沿 z 轴的加速力（包括重力）
TYPE_ACCELEROMETER_UNCALIBRATED 未经校准的加速度传感器（单位：m/s^2）	SensorEvent.values[0]	沿 x 轴的加速度，无偏差补偿
	SensorEvent.values[1]	沿 y 轴的加速度，无偏差补偿
	SensorEvent.values[2]	沿 z 轴的加速度，无偏差补偿
	SensorEvent.values[3]	沿 x 轴的加速度，有估算的偏差补偿
	SensorEvent.values[4]	沿 y 轴的加速度，有估算的偏差补偿
	SensorEvent.values[5]	沿 z 轴的加速度，有估算的偏差补偿
TYPE_GRAVITY 重力传感器（单位：m/s^2）	SensorEvent.values[0]	沿 x 轴的重力
	SensorEvent.values[1]	沿 y 轴的重力
	SensorEvent.values[2]	沿 z 轴的重力
TYPE_GYROSCOPE 陀螺仪传感器（单位：弧度/秒）	SensorEvent.values[0]	绕 x 轴的旋转速率
	SensorEvent.values[1]	绕 y 轴的旋转速率
	SensorEvent.values[2]	绕 z 轴的旋转速率

续表

传感器类型	传感器事件数据	说　明
TYPE_GYROSCOPE_UNCALIBRATED 未经校准的陀螺仪传感器（单位：弧度/秒）	SensorEvent.values[0]	绕 x 轴的旋转速率，无漂移补偿
	SensorEvent.values[1]	绕 y 轴的旋转速率，无漂移补偿
	SensorEvent.values[2]	绕 z 轴的旋转速率，无漂移补偿
	SensorEvent.values[3]	绕 x 轴的估算漂移
	SensorEvent.values[4]	绕 y 轴的估算漂移
	SensorEvent.values[5]	绕 z 轴的估算漂移
TYPE_LINEAR_ACCELERATION 线性加速度传感器（单位：m/s^2）	SensorEvent.values[0]	沿 x 轴的加速力（不包括重力）
	SensorEvent.values[1]	沿 y 轴的加速力（不包括重力）
	SensorEvent.values[2]	沿 z 轴的加速力（不包括重力）
TYPE_ROTATION_VECTOR 旋转矢量传感器	SensorEvent.values[0]	沿 x 轴的旋转矢量分量（$x \times \sin(\theta/2)$）
	SensorEvent.values[1]	沿 y 轴的旋转矢量分量（$y \times \sin(\theta/2)$）
	SensorEvent.values[2]	沿 z 轴的旋转矢量分量（$z \times \sin(\theta/2)$）
	SensorEvent.values[3]	旋转矢量的标量分量（$\cos(\theta/2)$）

8.4.3　实例工程：摇一摇比大小

本实例演示了摇动手机产生 0~9 随机数的聚会小游戏。单击"开始"按钮，摇动手机会自动生成数字。如果使用虚拟设备运行，需要打开虚拟传感器的控制界面，左右拖曳 y 轴移动的滑动条模拟上下摇晃效果（如图 8-8 所示）。

图 8-8　运行效果

1. 打开基础工程

打开"基础工程"文件夹中的"C0809"工程，该工程已经包含 MainActivity 类及布局的文件。

2. 主界面的 Activity

/java/com/vt/c0809/MainActivity.java

```java
39    //初始化传感器
40    private void init(){
41        mSensorManager = (SensorManager) getSystemService(SENSOR_SERVICE);//实例化传感器管理者
42        mSensor = mSensorManager.getDefaultSensor(Sensor.TYPE_ACCELEROMETER);//初始化加速度传感器
43        mSensorManager.registerListener(this, mSensor, SensorManager.SENSOR_DELAY_UI);
                                                                         //注册传感器监听器
44    }
45    @Override
46    public void onClick(View v) {
47        mNumTextView.setText("0");
48        if (mIsStart == true) {
49            mStartButton.setText("开始");
50            mIsStart = false;
51        } else {
52            mStartButton.setText("停止");
53            mIsStart = true;
54            //开启计时线程
55            mTimerThread = new Thread(this);
56            mTimerThread.start();
57        }
58    }
59    @Override
60    public void onSensorChanged(SensorEvent event) {
61        if (mIsStart == false) {
62            return;
63        }
64        float[] value = event.values;
65        mCurrentValue = getMod(value[0], value[1], value[2]);
66        //监测正向峰值
67        if (mMaxState == false) {
68            if (mCurrentValue >= mLastValue) {
69                mLastValue = mCurrentValue;
70            } else {
71                if (Math.abs(mCurrentValue - mLastValue) > mMotionRange) {
72                    mMaxState = true;
73                }
74            }
75        }
76        //监测反向峰值
77        if (mMaxState == true) {
78            if (mCurrentValue <= mLastValue) {
79                mLastValue = mCurrentValue;
80            } else {
81                if (Math.abs(mCurrentValue - mLastValue) > mMotionRange) {
82                    mCount++;
83                    mMaxState = false;
84                }
85            }
```

```
 86          }
 87      }
 88      @Override
 89      public void onAccuracyChanged(Sensor sensor, int accuracy) { }
 90      //向量求模
 91      public double getMod(float x, float y, float z) {
 92          return Math.sqrt(x * x + y * y + z * z);
 93      }
 94      @Override
 95      public void run() {
 96          //达到摇动次数结束循环
 97          while (mCount < mStopCount) {
 98              try {
 99                  Thread.sleep(1000);
100              } catch (InterruptedException e) {
101                  e.printStackTrace();
102              }
103              if(!mIsStart){
104                  return;
105              }
106          }
107          //摇动结束
108          mMainHandler.post(new Runnable() {
109              @Override
110              public void run() {
111                  mCount = 0;
112                  mIsStart = false;
113                  mStartButton.setText("开始");
114                  mNumTextView.setText(String.valueOf((int) (Math.random() * 10)));
115              }
116          });
117      }
118      @Override
119      protected void onDestroy() {
120          super.onDestroy();
121          mSensorManager.unregisterListener(this);
122      }
```

第 40～44 行初始化加速度传感器,并注册监听器。第 60～87 行重写 SensorEventListener 接口的 onSensorChanged(SensorEvent)方法,当检测到 y 轴的正向峰值到负向峰值后累加摇动次数 mCount。第 95～117 行重写 Runnable 接口的 run()方法,如果摇动次数 mCount 等于设定的停止摇动次数 mStopCount,则退出循环,并通过 mMainHandler.post(Runnable)方法对 UI 进行操作显示生成的随机数。第 119～122 行重写 onDestroy()方法,当 MainActivity 销毁时,注销传感器的监听器,以避免关闭 MainActivity 时报错。

8.4.4 位置类传感器

位置类传感器用于测量设备的物理位置或与附近物体的距离,包含旋转矢量传感器、地磁场传感器和近程传感器(如表 8-22 所示)。

表 8-22　Sensor 类的位置类传感器

传感器类型	传感器事件数据	说　　明
TYPE_GAME_ROTATION_VECTOR 游戏旋转矢量传感器	SensorEvent.values[0]	沿 x 轴的旋转矢量分量($x \times \sin(\theta/2)$)
	SensorEvent.values[1]	沿 y 轴的旋转矢量分量($y \times \sin(\theta/2)$)
	SensorEvent.values[2]	沿 z 轴的旋转矢量分量($z \times \sin(\theta/2)$)
TYPE_GEOMAGNETIC_ROTATION_VECTOR 地磁旋转矢量传感器	SensorEvent.values[0]	沿 x 轴的旋转矢量分量($x \times \sin(\theta/2)$)
	SensorEvent.values[1]	沿 y 轴的旋转矢量分量($y \times \sin(\theta/2)$)
	SensorEvent.values[2]	沿 z 轴的旋转矢量分量($z \times \sin(\theta/2)$)
TYPE_MAGNETIC_FIELD 地磁场传感器（单位：μT）	SensorEvent.values[0]	沿 x 轴的地磁场强度
	SensorEvent.values[1]	沿 y 轴的地磁场强度
	SensorEvent.values[2]	沿 z 轴的地磁场强度
TYPE_MAGNETIC_FIELD_UNCALIBRATED 未经校准的地磁场传感器（单位：μT）	SensorEvent.values[0]	沿 x 轴的地磁场强度（无硬铁校准）
	SensorEvent.values[1]	沿 y 轴的地磁场强度（无硬铁校准）
	SensorEvent.values[2]	沿 z 轴的地磁场强度（无硬铁校准）
	SensorEvent.values[3]	沿 x 轴的铁偏差估算
	SensorEvent.values[4]	沿 y 轴的铁偏差估算
	SensorEvent.values[5]	沿 z 轴的铁偏差估算
TYPE_PROXIMITY 近程传感器（单位：cm）	SensorEvent.values[0]	与物体的距离

8.4.5　实例工程：指南针

本实例演示了指南针及手机姿态的数据，单击虚拟设备右侧的旋转按钮可以改变手机的方向，观察方向和数据的变化（如图 8-9 所示）。手机方位的数据还可以用于拍摄照片和录制视频时保存文件的方向，解决之前实例工程中横向拍摄保存的照片方向错误的问题。

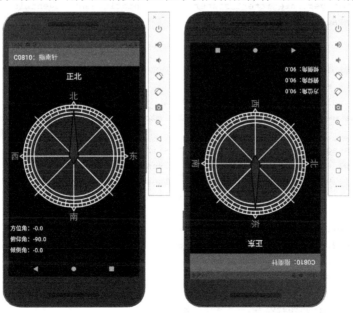

图 8-9　运行效果

1. 打开基础工程

打开"基础工程"文件夹中的"C0810"工程，该工程已经包含图片资源、MainActivity 类及布局的文件。

2. 主界面的 Activity

```
/java/com/vt/c0810/MainActivity.java
39    //初始化
40    private void init(){
41        mSensorManager = (SensorManager) getSystemService(Context.SENSOR_SERVICE);
                                                                    //实例化传感器管理者
42        mAccelerometer = mSensorManager.getDefaultSensor(Sensor.TYPE_ACCELEROMETER);
                                                                    //初始化加速度传感器
43        mMagnetic = mSensorManager.getDefaultSensor(Sensor.TYPE_MAGNETIC_FIELD);
                                                                    //初始化地磁场传感器
44        calculateOrientation();//计算方向
45    }
46    @Override
47    protected void onResume() {
48        mSensorManager.registerListener(new OrientationSensorEventListener(), mAccelerometer,
          SensorManager.SENSOR_DELAY_NORMAL);
49        mSensorManager.registerListener(new OrientationSensorEventListener(), mMagnetic,
          SensorManager.SENSOR_DELAY_NORMAL);
50        super.onResume();
51    }
52    @Override
53    protected void onPause() {
54        mSensorManager.unregisterListener(new OrientationSensorEventListener());
55        super.onPause();
56    }
57    //计算方向
58    private void calculateOrientation() {
59        float[] R = new float[9];//旋转矩阵
60        float[] values = new float[3];//方位数组
61        SensorManager.getRotationMatrix(R, null, mAccelerometerValues, mMagneticFieldValues);
                                                                    //获取旋转矩阵
62        SensorManager.getOrientation(R, values);//获取方位数组
63        values[0] = (float) Math.toDegrees(values[0]);
64        values[1] = (float) Math.toDegrees(values[1]);
65        values[2] = (float) Math.toDegrees(values[2]);
66        //手机姿态
67        mTextView1.setText("方位角: " + values[0]);
68        mTextView2.setText("俯仰角: " + values[1]);
69        mTextView3.setText("倾侧角: " + values[2]);
70        //手机正前方水平方向
71        if (values[0] >= -5 && values[0] < 5) {
72            mOrientationTextView.setText("正北");
73        } else if (values[0] >= 5 && values[0] < 85) {
74            mOrientationTextView.setText("东北");
```

```
75        } else if (values[0] >= 85 && values[0] <= 95) {
76            mOrientationTextView.setText("正东");
77        } else if (values[0] >= 95 && values[0] < 175) {
78            mOrientationTextView.setText("东南");
79        } else if ((values[0] >= 175 && values[0] <= 180) || (values[0]) >= -180 && values[0] < -175) {
80            mOrientationTextView.setText("正南");
81        } else if (values[0] >= -175 && values[0] < -95) {
82            mOrientationTextView.setText("西南");
83        } else if (values[0] >= -95 && values[0] < -85) {
84            mOrientationTextView.setText("正西");
85        } else if (values[0] >= -85 && values[0] < -5) {
86            mOrientationTextView.setText("西北");
87        }
88        //转换指南针指针方向
89        int degree = -(int)values[0];
90        if (degree - mCurrentDegree > 180) {
91            degree -= 360;
92        } else if (degree - mCurrentDegree < -180) {
93            degree += 360;
94        }
95        //改变值大于2° 改变指针方向
96        if (Math.abs(degree - mCurrentDegree) > 2) {
97            RotateAnimation ra = new RotateAnimation(mCurrentDegree, degree,
98                    Animation.RELATIVE_TO_SELF, 0.5f, Animation.RELATIVE_TO_SELF, 0.5f);
99            ra.setDuration(200);//旋转持续时间
100           ra.setFillAfter(true);//旋转结束后停留
101           mPointerImageView.startAnimation(ra);//开始旋转动画
102           mCurrentDegree = degree;//更新当前指针角度
103       }
104  }
105  //传感器事件监听器
106  class OrientationSensorEventListener implements SensorEventListener {
107      @Override
108      public void onSensorChanged(SensorEvent event) {
109          if (event.sensor.getType() == Sensor.TYPE_ACCELEROMETER) {
110              mAccelerometerValues = event.values;
111          }
112          if (event.sensor.getType() == Sensor.TYPE_MAGNETIC_FIELD) {
113              mMagneticFieldValues = event.values;
114          }
115          calculateOrientation();
116      }
117      @Override
118      public void onAccuracyChanged(Sensor sensor, int accuracy) { }
119  }
```

第 40~45 行初始化加速度传感器和地磁场传感器，并调用 calculateOrientation()方法计算手机指向的方向。第 47~51 行重写 onResume()方法，当 MainActivity 显示到前台后，注册加速度传感器和地磁场传感器的监听器。第 53~56 行重写 onPause()方法，注销传感器的监听器。第 106~119 行自定义实现 SensorEventListener 接口的 OrientationSensorEventListener 类，重写 onSensorChanged(SensorEvent)方法获取加速度传感器和地磁场传感器的数据，最后调用 calculateOrientation()方法计算并显示方位。

8.4.6 环境类传感器

环境类传感器用于测量环境气温、气压、照度和湿度等参数，包含温度传感器、光度传感器、气压传感器和湿度传感器（如表 8-23 所示）。

表 8-23　Sensor 类的环境类传感器

传感器类型	传感器事件数据	说　　明
TYPE_AMBIENT_TEMPERATURE 温度传感器（单位：摄氏度）	SensorEvent.values[0]	温度
TYPE_LIGHT 光度传感器（单位：勒克斯）	SensorEvent.values[0]	照度
TYPE_PRESSURE 气压传感器（单位：百帕斯卡）	SensorEvent.values[0]	空气压力
TYPE_RELATIVE_HUMIDITY 湿度传感器（单位：百分比）	SensorEvent.values[0]	相对湿度

8.4.7　实例工程：光照计和气压计

本实例演示了光照计和气压计，并且能够根据光照强度改变屏幕亮度。可以在虚拟设备中设置虚拟传感器的参数，观察模拟效果（如图 8-10 所示）。海拔高度是通过气压计算得到的，由于温度和风速会对气压产生影响，因此这种方式计算海拔高度的精度不高。

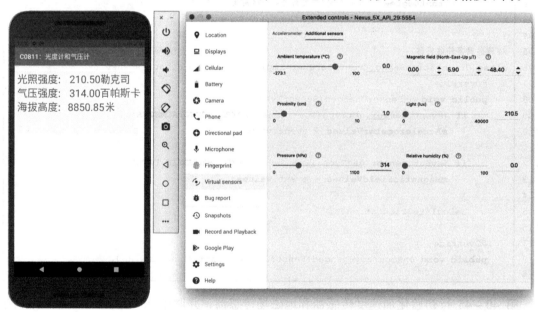

图 8-10　运行效果

1．打开基础工程

打开"基础工程"文件夹中的"C0811"工程，该工程已经包含 MainActivity 类及布局的文件。

2．主界面的 Activity

```
/java/com/vt/c0811/MainActivity.java
33    //初始化
34    private void init(){
35        mWindow = this.getWindow();
36        mSensorManager = (SensorManager) getSystemService(Context.SENSOR_SERVICE);
37        mLightSensor = mSensorManager.getDefaultSensor(Sensor.TYPE_LIGHT);//光度传感器
38        mPressureSensor = mSensorManager.getDefaultSensor(Sensor.TYPE_PRESSURE);//气压传感器
39        //注册光度传感器
40        mSensorManager.registerListener(new SensorEventListener() {
41            @Override
42            public void onSensorChanged(SensorEvent event) {
43                float lux = event.values[0];//获取照度
44                mLightTextView.setText(new DecimalFormat("#.00").format(lux));
45                //根据照度调整屏幕亮度
46                WindowManager.LayoutParams lp = mWindow.getAttributes();
47                lp.screenBrightness = lux / 500;
48                mWindow.setAttributes(lp);
49            }
50            @Override
51            public void onAccuracyChanged(Sensor sensor, int accuracy) { }
52        }, mLightSensor, SensorManager.SENSOR_DELAY_NORMAL);
53        //注册气压传感器
54        mSensorManager.registerListener(new SensorEventListener() {
55            @Override
56            public void onSensorChanged(SensorEvent event) {
57                double pa = event.values[0];//获取大气压力
58                mPressureTextView.setText(new DecimalFormat("#.00").format(pa));
59                mAltitudeTextView.setText(new DecimalFormat("#.00").format(altitude(pa)));
60            }
61            @Override
62            public void onAccuracyChanged(Sensor sensor, int accuracy) { }
63        }, mPressureSensor, SensorManager.SENSOR_DELAY_NORMAL);
64    }
65    //通过气压计算海拔高度
66    private double altitude(double pa){
67        float p0 = 1013.21f;//海平面大气压
68        return 44300*(1- Math.pow(pa/p0,1/5.256));
69    }
```

第 40～52 行注册光度传感器的监听器，并重写 onSensorChanged(SensorEvent)方法，格式化显示照度和根据光照度改变屏幕亮度。第 54～63 行注册气压传感器，并重写 onSensorChanged(SensorEvent)方法，显示气压和通过气压计算的高度。第 66～69 行 altitude(double)通过气压计算海拔高度，这种方式计算的海拔高度受温度和风速的影响较大。

8.5 位置服务

8.5.1 位置服务概述

LocationManager 类（android.location.LocationManager）用于系统位置服务的访问管理（如表 8-24 所示），系统位置服务定期更新位置或进入指定地理位置发送通知。定位方式有 LocationManager.GPS_PROVIDER（GPS 获取定位信息）、LocationManager.NETWORK_PROVIDER（网络获取定位信息）和 LocationManager PASSIVE_PROVIDER（其他 App 提供定位信息）三种。

表 8-24 LocationManager 类的常用方法

类型和修饰符	方 法
boolean	addNmeaListener(OnNmeaMessageListener listener, Handler handler) 添加一个 NMEA 监听器
void	addProximityAlert(double latitude, double longitude, float radius, long expiration, PendingIntent intent) 添加接近警报，根据位置（纬度、经度）和半径指定的位置设置接近警报
List<String>	getAllProviders() 获取所有位置提供者的名称列表
void	getCurrentLocation(String provider, CancellationSignal cancellationSignal, Executor executor, Consumer<Location> consumer) 异步获取当前位置
String	getGnssHardwareModelName() 获取 GNSS（全球导航卫星系统）硬件驱动程序的型号名称
Location	getLastKnownLocation(String provider) 获取最后一个已知位置，如果没有则返回 null
boolean	isLocationEnabled() 判断位置服务是否启用
void	removeNmeaListener(OnNmeaMessageListener listener) 移除 NMEA 监听器
void	removeProximityAlert(PendingIntent intent) 移除接近警报
void	removeUpdates(LocationListener listener) 移除指定的 LocationListener 的位置更新
void	requestLocationUpdates(String provider, long minTimeMs, float minDistanceM, LocationListener listener) 注册位置提供者的位置更新监听器

Location 类（android.location.Location）用于存储地理位置（如表 8-25 所示），包含纬度、经度、时间戳和其他信息，其他信息是可选项。

表 8-25 Location 类的常用方法

类型和修饰符	方 法
	Location(String provider) 构造方法
float	bearingTo(Location dest) 沿此位置与给定位置之间的最短路径行驶时，返回近似初始方位角

续表

类型和修饰符	方　　法
static String	convert(double coordinate, int outputType) 将坐标转换为字符串形式
static double	convert(String coordinate) 将字符串形式的坐标转换为双精度型
static void	distanceBetween(double startLatitude, double startLongitude, double endLatitude, double endLongitude, float[] results) 计算两个位置之间的近似距离（以 m 为单位），以及两个位置之间最短路径的初始和最终方位
float	distanceTo(Location dest) 返回此位置和给定位置之间的近似距离（以 m 为单位）
float	getAccuracy() 获取精度（以 m 为单位）
double	getAltitude() 获取海拔高度
float	getBearing() 获取方位角（以度为单位）
float	getBearingAccuracyDegrees() 获取方位角精度（以度为单位）
double	getLatitude() 获取纬度（以度为单位）
double	getLongitude() 获取经度（以度为单位）
float	getSpeed() 获取速度（以 m/s 为单位）
void	reset() 清除位置的内容
void	set(Location l) 将位置的内容设置为给定位置的值
void	setAltitude(double altitude) 设置高度（以 m 为单位）
void	setBearing(float bearing) 设置方位角（以度为单位）
void	setLatitude(double latitude) 设置纬度（以度为单位）
void	setLongitude(double longitude) 设置经度（以度为单位）
void	setSpeed(float speed) 设置速度（以 m/s 为单位）

提示：反向地理编码

反向地理编码，又称为逆向地理编码，是根据 GPS 坐标反向查询所在的街道名称或所在建筑名称，在社交 App 中最为常见。该功能需要大量数据采集和定期数据更新，因此推荐使用第三方提供的服务实现该功能。国内推荐使用百度或高德开放平台 SDK、API，国外可以使用 Google Map 的 API。

8.5.2 实例工程：获取经纬度坐标

本实例演示了使用 GPS 卫星和网络进行定位（如图 8-11 所示），两种定位方式各有利弊。卫星定位精度较高，但是室内往往搜索不到卫星信号，无法使用卫星定位。网络定位精度较低，而且通常无法获取高度、速度和方位角，优势在于耗电量小。

1. 打开基础工程

打开"基础工程"文件夹中的"C0812"工程，该工程已经包含 MainActivity 类及布局的文件。在 AndroidManifest.xml 文件中添加 <uses-permission android:name="android.permission.ACCESS_FINE_LOCATION"/>权限。

图 8-11 运行效果

2. 主界面的 Activity

```
/java/com/vt/c0812/MainActivity.java
51  //初始化
52  @SuppressLint("MissingPermission")
53  private void init() {
54      mRadioGroup = findViewById(R.id.radio_group);
55      mRadioGroup.setOnCheckedChangeListener(this);
56      mLocationTextView = findViewById(R.id.location_text_view);
57      //获取定位服务
58      mLocationManager = (LocationManager) getSystemService(Context.LOCATION_SERVICE);
59      if (!isGpsAble(mLocationManager)) {
60          Toast.makeText(MainActivity.this, "请打开 GPS", Toast.LENGTH_SHORT).show();
61          openGps();
62      }
63      //默认使用 GPS 卫星进行定位
64      mLocation = mLocationManager.getLastKnownLocation(LocationManager.GPS_PROVIDER);
65      displayLocationData(mLocation);
66      //设置间隔 2s 请求一次 GPS 卫星定位信息
67      mLocationManager.requestLocationUpdates(LocationManager.GPS_PROVIDER, 2000, 8, new LocationListener() {
68          @Override
69          public void onLocationChanged(Location location) {
70              displayLocationData(location);
71          }
72          @Override
73          public void onStatusChanged(String provider, int status, Bundle extras) { }
74          @Override
75          public void onProviderEnabled(String provider) { }
76          @Override
77          public void onProviderDisabled(String provider) {
78              displayLocationData(null);
```

```
79         }
80      });
81  }
82  //显示定位数据
83  private void displayLocationData(Location location) {
84      if (location != null) {
85          StringBuilder sb = new StringBuilder();
86          sb.append("经度: " + location.getLongitude() + "\n");
87          sb.append("纬度: " + location.getLatitude() + "\n");
88          sb.append("高度: " + location.getAltitude() + "\n");
89          sb.append("速度: " + location.getSpeed() + "\n");
90          sb.append("方位角: " + location.getBearing() + "\n");
91          sb.append("定位精度: " + location.getAccuracy() + "\n");
92          mLocationTextView.setText(sb.toString());
93      } else {
94          mLocationTextView.setText("无法获取定位数据");
95      }
96  }
```

第 64 行获取最后的定位数据。第 65 行调用 displayLocationData(Location)方法显示定位数据。第 67～80 行每隔 2000ms 请求一次定位数据的更新，如果定位数据有更新，则通过重写的 onLocationChanged(Location)方法获取更新，并调用 displayLocationData(Location)方法显示定位数据。第 83～96 行 displayLocationData(Location)方法根据传递的定位数据，经解析后显示在 mLocationTextView 对象中，如果获取不到定位数据，则显示无法获取的提示。

8.6 习　　题

1. 使用 Camera2 实现定时拍照的功能。
2. 使用运动传感器实现步数统计的功能。
3. 使用 Camera2 拍照时通过 GPS 定位功能获取坐标，并将其保存在照片文件的详细信息中。

第 9 章 HTTP 网络通信

HTTP 网络通信是实现起来最为便捷的网络通信方式，特别适合对低延迟要求不高，且与服务器端非持续连接的数据通信。HTTP 网络通信需要客户端发送请求，服务器端才能发送数据。HTTP 除了不适合移动设备作为服务器端，还特别不适合应用于低延迟的网络游戏和实时音视频传输。

9.1 HttpURLConnection 类

HttpURLConnection 类（java.net.HttpURLConnection）使用 HTTP 进行数据传输（如表 9-1 所示），GET 和 POST 方法是 HTTP 使用较频繁的方法。通过 HTTP 向服务器端发送请求后，服务器端会返回状态码用于快速判断响应结果（如表 9-2 所示）。

表 9-1　HttpURLConnection 类的常用方法

类型和修饰符	方　　法
	HttpURLConnection(URL u) 构造方法
void	addRequestProperty(String key, String value) 添加请求属性
abstract void	connect() 打开连接
abstract void	disconnect() 断开连接
String	getHeaderField(int n) 获取消息头的字段数据
long	getHeaderFieldDate(String name, long Default) 获取消息头的日期字段数据
String	getHeaderFieldKey(int n) 获取消息头的字段
String	getRequestMethod() 获取请求方式
int	getResponseCode() 获取 HTTP 响应消息的状态代码
String	getResponseMessage() 获取 HTTP 的响应消息

续表

类型和修饰符	方　法
void	setConnectTimeout(int timeout) 设置连接的超时时间（以 ms 为单位）
void	setReadTimeout(int timeout) 设置下载数据的超时时间（以 ms 为单位）
void	setRequestMethod(String method) 设置请求方法（GET、POST、HEAD、OPTIONS、PUT、DELETE 或 TRACE）

表 9-2　HTTP 的常用状态码

状态码	HttpURLConnection 类的常量	说　明
200	HTTP_OK	请求已成功，请求的响应头或数据体将随之返回
400	HTTP_BAD_REQUEST	请求无效
404	HTTP_NOT_FOUND	请求失败
408	HTTP_CLIENT_TIMEOUT	请求超时
500	HTTP_INTERNAL_ERROR	服务器内部错误，多源于服务器端的源代码出现错误
505	HTTP_VERSION	服务器版本不支持

> **提示：HTTP 和 HTTPS**
>
> HTTP（Hyper Text Transfer Protocol）是从 Web 服务器传输超文本到本地浏览器的传输协议，基于 TCP/IP 来传递数据（HTML 文件、图片文件、查询结果等）。浏览器作为客户端通过 URL 向服务器端的 Web 服务器发送所有请求，Web 服务器根据接收到的请求，向客户端发送响应信息。
>
> HTTPS（Hyper Text Transfer Protocol over Secure Socket Layer）是由 HTTP 加上 TLS/SSL 协议构建的可进行加密传输、身份认证的网络协议，主要通过数字证书、加密算法、非对称密钥等技术完成互联网数据传输加密，实现互联网传输安全保护。在相同的网络环境下，HTTPS 会使页面的加载时间延长 50%，且增加 10% 到 20% 的耗电量。

9.2　实例工程：加载网络图片（带缓存）

本实例演示了使用 HttpURLConnection 类下载图片并进行本地缓存，以及清除本地缓存（如图 9-1 所示）。单击"加载图片"按钮，开始下载图片，图片下载完成后保存在本地缓存文件夹中，然后显示在 ImageView 控件中。单击"清除缓存"按钮，清除本地缓存，在 ImageView 控件中显示 R.mipmap.img_error_m 资源图片。

1. 打开基础工程

打开"基础工程"文件夹中的"C0901"工程，该工程已经包含 MainActivity、FileHelper 类及布局的文件。

图 9-1 运行效果

2. 设置权限

/manifests/AndroidManifest.xml	
02	`<manifest xmlns:android="http://schemas.android.com/apk/res/android" package="com.vt.c0901">`
03	`    <uses-permission android:name="android.permission.INTERNET"/>`
04	`    <uses-permission android:name="android.permission.ACCESS_NETWORK_STATE"/>`
05	`    <application`
06	`        android:allowBackup="true"`
07	`        android:icon="@mipmap/ic_launcher"`
08	`        android:label="@string/app_name"`
09	`        android:roundIcon="@mipmap/ic_launcher_round"`
10	`        android:supportsRtl="true"`
11	`        android:theme="@style/AppTheme"`
12	`        android:usesCleartextTraffic="true">`

第 03 行 android.permission.INTERNET 是访问互联网的权限。第 04 行 android.permission.ACCESS_NETWORK_STATE 是访问网络状态的权限。第 12 行 android:usesCleartextTraffic="true"表示允许使用明文访问网络，Android 9 或更高版本需要将该属性设置为 true，否则只能访问 HTTPS 地址，访问 HTTP 地址时会报错。

3. Http 类

在 "/java/com/vt/c0901/" 文件夹中，新建 "Http.java" 文件，用于添加下载文件的静态方法。

/java/com/vt/c0901/Http.java	
10	`    private static final String EXCEPTION_FAILURE = "请求URL失败";`
11	`    public static File downloadFile(String path, String cachePath) throws Exception {`
12	`        //本地缓存文件`
13	`        File storageDir = new File(cachePath);//判断缓存文件夹是否存在`
14	`        if (!storageDir.exists()) {`

```
15            storageDir.mkdirs();
16        }
17        String[] temp = path.split("\\/");//折分路径字符串
18        String fileName = temp[temp.length - 1];//获取文件名
19        File cacheFile = new File(storageDir, fileName);//缓存文件
20        //下载请求
21        URL url = new URL(path);//将字符串格式的路径转为 URL
22        HttpURLConnection conn = (HttpURLConnection) url.openConnection();
23        conn.setConnectTimeout(5000);//设置连接超时时间
24        conn.setReadTimeout(5000);//设置下载超时时间
25        conn.setRequestMethod("GET");//设置请求类型为 Get 类型
26        if (conn.getResponseCode() != 200) {//判断请求 URL 是否成功
27            throw new RuntimeException(EXCEPTION_FAILURE);
28        }
29        //如果有本地缓存文件，则不进行下载而直接使用
30        long contentLength = conn.getContentLengthLong();//文件数据大小
31        if (cacheFile.exists()) {//判断是否缓存
32            //如果缓存文件与下载文件大小相等，则直接返回缓存文件
33            if(cacheFile.length() == contentLength)
34                return cacheFile;
35        }
36        //下载文件
37        InputStream is = conn.getInputStream();//获取输入数据流
38        FileOutputStream fileOutputStream = null;//文件输出数据流
39        if (is != null) {
40            fileOutputStream = new FileOutputStream(cacheFile);//指定文件保存路径
41            byte[] buf = new byte[1024];
42            int ch;
43            while ((ch = is.read(buf)) != -1) {
44                fileOutputStream.write(buf, 0, ch);//将获取到的数据流写入文件中
45            }
46        }
47        //释放资源
48        if (fileOutputStream != null) {
49            fileOutputStream.flush();
50            fileOutputStream.close();
51        }
52        conn.disconnect();
53        return cacheFile;
54    }
```

第 13~18 行获取本地缓存文件，如果缓存目录不存在，则创建缓存目录。第 21~28 行通过 URL 发送请求，如果请求失败，则抛出异常。第 30~35 行如果缓存文件存在且文件大小相同，则直接加载缓存文件。第 37~46 行通过读取数据流，将文件数据写入缓存文件。

4．ImageLoader 类

在 "/java/com/vt/c0901/" 文件夹中，新建 "ImageLoader.java" 文件，用于使用独立线程下载图片并显示在指定的 ImageView 控件中。

```
/java/com/vt/c0901/ImageLoader.java
09  public class ImageLoader {
10      private static final String EXCEPTION_FAILURE = "请求URL失败";
11      private static final int MSG_OK = 0;
12      private static final int MSG_ERROR = 1;
13      private ImageView mImageView;
14      private File mDownloadFile;
15      //使用独立线程加载图片
16      public void displayImage(ImageView imageView, final String imageUrl, final String cacheDir) {
17          mImageView = imageView;
18          new Thread(new Runnable() {
19              @Override
20              public void run() {
21                  try {
22                      mDownloadFile = downloadFile(imageUrl, cacheDir);
23                      if (mDownloadFile.exists()) {
24                          mHandler.obtainMessage(MSG_OK).sendToTarget();
25                      }
26                  } catch (Exception e) {
27                      e.printStackTrace();
28                      mHandler.obtainMessage(MSG_ERROR).sendToTarget();
29                  }
30              }
31          }).start();
32      }
33      //更新UI
34      private Handler mHandler = new Handler() {
35          public void handleMessage(Message msg) {
36              switch (msg.what) {
37                  case MSG_OK://下载成功
38                      mImageView.setImageURI(Uri.fromFile(mDownloadFile));
39                      break;
40                  case MSG_ERROR://下载出错
41                      //根据控件宽度显示不同的错误图片
42                      if(mImageView.getWidth()>900){
43                          mImageView.setImageResource(R.mipmap.img_error_h);
44                      }else if(mImageView.getWidth()>300){
45                          mImageView.setImageResource(R.mipmap.img_error_m);
46                      }else{
47                          mImageView.setImageResource(R.mipmap.img_error_s);
48                      }
49                      break;
50              }
51          }
52      };
53  }
```

第 18~31 行开启新线程加载图片，如果加载成功，使用 mHandler.obtainMessage(MSG_OK).sendToTarget()发送成功的信息。第 34~52 行 mHandler 对象接收发送的信息，通过 handleMessage(Message)方法根据发送信息的内容(msg.what)进行分类处理。

5. 主界面的 Activity

```
/java/com/vt/c0901/MainActivity.java
19    mUrl = "http://www.weiju2014.com/teachol/android/IMG225.jpg";
20    //初始化加载图片
21    mImageView = findViewById(R.id.image_view);
22    mImageLoader = new ImageLoader();
23    //加载图片
24    Button displayButton = findViewById(R.id.display_button);
25    displayButton.setOnClickListener(new View.OnClickListener() {
26        @Override
27        public void onClick(View v) {
28            mImageLoader.displayImage(mImageView, mUrl,getExternalCacheDir().getAbsolutePath()
      + "/image");
29        }
30    });
31    //清除缓存
32    Button cleanCacheButton = findViewById(R.id.clean_cache_button);
33    cleanCacheButton.setOnClickListener(new View.OnClickListener() {
34        @Override
35        public void onClick(View v) {
36            FileHelper.delete(getExternalCacheDir().getAbsolutePath() + "/image");
37            mImageView.setImageResource(R.mipmap.img_error_m);
38        }
39    });
```

第 28 行使用 mImageLoader 加载图片,如果加载成功,则将图片显示在 mImageView 控件中。第 36 行调用 FileHelper.delete(String)静态方法,删除缓存文件夹中的所有文件。

扩展实例:下载提示

在下载时还可以使用等待提示或下载进度对用户进行提示(如图 9-2 所示),C0902 工程和 C0903 工程分别使用两种效果提示下载图片,由于篇幅所限,可打开工程文件自行查看,其中包含详细的注释。

图 9-2 带下载提示和下载进度的效果

9.3 实例工程：发布动态（POST 方式）

本实例演示了使用 HttpURLConnection 类发送带附件的 POST 请求，服务器端检验后返回成功值并关闭 Activity（如图 9-3 所示）。单击"动态列表"按钮，启动 DailyListActivity。单击+按钮，使用系统提供的功能选择图片，选择图片后对图片进行压缩并保存在本地缓存文件夹中。这样处理不但可以避免 OOM 异常的出现，还能减小上传的数据量。单击"发布"按钮，将控件输入的昵称和内容及选择后缓存在本地的缓存图片合成为 POST 数据流进行发送。

图 9-3　运行效果

1. 打开基础工程

打开"基础工程"文件夹中的"C0904"工程，该工程已经包含 MainActivity、DailyAddActivity、Http、CompressImage、Util 类及布局的文件。CompressImage 类和 Util 类预先添加了一些功能性的方法，由于篇幅限制，详细的代码可以在工程文件中查看，其中有详细的注释帮助理解代码。

CompressImage 类除了构造方法，还包含 3 个方法。compress(ContentResolver contentResolver, Uri uri)方法用于压缩位图，getSampleSize(Bitmap bitmap)方法用于获取缩放采样，compressByQuality(Bitmap image) 方法用于压缩质量。

Util 类添加了 3 个方法。UriToString(ContentResolver contentResolver, Uri fileUrl)方法用于将 URI 转为 String；getImageOrientation(Context context, Uri photoUri)方法用于获取图片旋转角度；saveCacheBitmap(String cachePath, String filename, Bitmap bitmap)方法用于保存缓存图片。

2. Http 类

/java/com/vt/c0904/Http.java

```
71   //post 请求
72   public static String post(String requestURL, Map<String, File> files, Map<String,
     String> params) throws Exception {
73       URL url = new URL(requestURL);
74       String BOUNDARY = UUID.randomUUID().toString();//边界标识（随机生成）
75       String PREFIX = "--";//前缀字符串
76       String LINE_END = "\r\n";//换行字符串
77       String CONTENT_TYPE = "multipart/form-data";//内容类型
78       //创建连接
79       HttpURLConnection conn = (HttpURLConnection) url.openConnection();
80       conn.setConnectTimeout(CONNECT_TIME_OUT);
81       conn.setReadTimeout(READ_TIME_OUT);
82       conn.setDoInput(true);//允许输入数据流
83       conn.setDoOutput(true);//允许输出数据流
84       conn.setUseCaches(false);//不允许使用缓存
85       conn.setRequestMethod("POST");//请求方式
86       conn.setRequestProperty("Charset", CHARSET);//设置编码
87       conn.setRequestProperty("connection", "keep-alive");
88       conn.setRequestProperty("Content-Type", CONTENT_TYPE + ";boundary=" + BOUNDARY);
89       //设置数据流
90       OutputStream outputSteam = conn.getOutputStream();
91       DataOutputStream dos = new DataOutputStream(outputSteam);
92       StringBuffer sb = new StringBuffer();
93       //添加参数
94       for (Map.Entry<String, String> entry : params.entrySet()) {
95           sb.append(PREFIX);
96           sb.append(BOUNDARY);
97           sb.append(LINE_END);
98           sb.append("Content-Disposition: form-data; name=\"" + entry.getKey() + "\"" + LINE_END);
99           sb.append("Content-Type: text/plain; charset=" + CHARSET + LINE_END);
100          sb.append("Content-Transfer-Encoding: 8bit" + LINE_END);
101          sb.append(LINE_END);
102          sb.append(entry.getValue());
103          sb.append(LINE_END);
104          dos.write(sb.toString().getBytes());
105          Log.e(TAG, entry.getKey() + ": " + entry.getValue());
106      }
107      //添加文件
108      if (files != null) {
109          for (Map.Entry<String, File> file : files.entrySet()) {
110              sb.append(PREFIX);
111              sb.append(BOUNDARY);
112              sb.append(LINE_END);
113              sb.append("Content-Disposition: form-data; name=\"uploadinput[]\"; filename=\"" +
     file.getKey() + "\"" + LINE_END);//uploadinput[]是服务器端用于接收图片文件的变量数组, filename是文件名
114              sb.append("Content-Type: multipart/form-data; charset=" + CHARSET + LINE_END);
115              sb.append(LINE_END);
```

```java
116         dos.write(sb.toString().getBytes());
117         Log.e(TAG, "file:" + file.getKey());
118         //写入图片文件数据
119         InputStream is = new FileInputStream(file.getValue());
120         byte[] buffer = new byte[1024];
121         int len;
122         while ((len = is.read(buffer)) != -1) {
123             dos.write(buffer, 0, len);
124         }
125         is.close();
126         dos.write(LINE_END.getBytes());
127     }
128 }
129 //写入结束标志
130 byte[] end_data = (PREFIX + BOUNDARY + PREFIX + LINE_END).getBytes();
131 dos.write(end_data);
132 //发送数据流
133 dos.flush();
134 //获取服务器端的响应码,200表示发送成功
135 if (conn.getResponseCode() == 200) {
136     BufferedReader br = new BufferedReader(new InputStreamReader(conn.getInputStream()));
137     String state = br.readLine();//读取服务器端返回的数据
138     Log.e(TAG, "response getInputStream:" + state);
139     Log.e(TAG, "response Message:" + conn.getResponseMessage());
140     return state;
141 } else {
142     throw new RuntimeException(FAILURE);
143 }
144 }
```

第 79～88 行设置 POST 类型的 HTTP 请求。第 94～106 行遍历 params 将所有上传的参数写入请求的数据流中。第 108～128 行遍历 files 将所有上传的附件数据写入请求的数据流中。第 135～143 行根据返回的响应码进行相应处理,如果响应码等于 200,返回响应数据,否则抛出异常。

3. 发布动态的 Activity

/java/com/vt/c0904/DailyActivity.java
```java
70  //单击事件
71  @Override
72  public void onClick(View v) {
73      switch (v.getId()) {
74          case R.id.image_view://选择图片
75              i = 0;
76              Intent intent = new Intent(Intent.ACTION_PICK);
77              intent.setType("image/*");
78              startActivityForResult(intent, RESULT_CANCELED);
79              break;
80          case R.id.send_button://发布动态
81              if (!mNameEditText.getText().toString().equals("")
```

```
82                      &&!mContentEditText.getText().toString().equals("")) {
83                  //设置上传地址
84                  String url = "http://www.weiju2014.com/teachol/android/DailyAdd.php";
85                  //post 参数
86                  Map<String, String> params = new HashMap<>();
87                  params.put("name", mNameEditText.getText().toString());
88                  params.put("content", mContentEditText.getText().toString());
                    //将选择的图片存储在数组中
89                  ArrayList<String> selectFilePath = new ArrayList<>();
90                  for (String path : mImagePath) {
91                      if (!path.equals("null")) {
92                          selectFilePath.add(path);
93                      }
94                  }
95                  String[] uploadFilePath = new String[selectFilePath.size()];
96                  selectFilePath.toArray(uploadFilePath);
97                  //post 上传
98                  pd = ProgressDialog.show(mContext, "", "正在发布中……", false, false);
99                  mHttpPostThread = new HttpPostThread();
100                 mHttpPostThread.url = url;
101                 mHttpPostThread.params = params;
102                 mHttpPostThread.filePath = uploadFilePath;
103                 mHttpPostThread.start();
104                 hideSoftInput();//关闭软键盘
105             } else {
106                 Toast.makeText(this, "亲,写点什么吧!", Toast.LENGTH_SHORT).show();
107             }
108             break;
109         }
110 }
158 //上传动态数据线程
159 public class HttpPostThread extends Thread {
160     private String url;
161     private Map<String, String> params;
162     private String[] filePath;
163     @Override
164     public void run() {
165         //压缩并缓存图片
166         File cacheFile;//缓存文件
167         Map<String, File> postFiles = new HashMap<>();//上传图片文件
168         for (int i = 0; i < filePath.length; i++) {
169             //对大图片进行压缩和旋转
170             ContentResolver cr = mContext.getContentResolver();
171             CompressImage ci = new CompressImage(cacheDir, 500 * 1024);
172             Bitmap bitmap = null;
173             try {
174                 bitmap = ci.compress(cr,mImageUri[i]);
175             } catch (IOException e) {
176                 e.printStackTrace();
177             }
```

```
178                bitmap = Util.rotateBitmapByDegree(bitmap, mImageDegrees[i]);
179                cacheFile = Util.saveCacheBitmap(cacheDir,"IMG" + i + ".jpg", bitmap);
180                //添加上传图片文件
181                postFiles.put(cacheFile.getName(), cacheFile);
182            }
183            //post 方式上传
184            try {
185                if (!Http.post(url, postFiles, params).equals(Http.FAILURE)) {
186                    mHandler.obtainMessage(MSG_HTTP_SUCCESS).sendToTarget();
187                } else {
188                    mHandler.obtainMessage(MSG_HTTP_FAILURE).sendToTarget();
189                }
190            } catch (Exception e) {
191                e.printStackTrace();
192            }
193        }
194    }
```

第 76～78 行调用系统功能选择图片。第 82～103 行开启独立线程发送 POST 请求发布动态。第 159～194 行 HttpPostThread 类继承自 Thread 类，实现压缩附件文件并发送 POST 请求发布动态。

4．测试运行

运行工程后，单击"发布动态"按钮启动 DailyActivity，输入昵称和内容，再选择图片，然后单击"发布"按钮。发送成功后关闭 DailyActivity，并在"Logcat"窗口中显示发送的内容及响应码和响应数据（如图 9-4 所示）。

图 9-4 "Logcat"窗口输出的数据

扩展实例：绝对路径访问本地图片

C0905 工程使用绝对路径的旧方式访问本地图片，CompressImage 类的 compress(String) 方法使用了绝对路径，Android 10（API level 29）以上需要添加<application android:requestLegacyExternalStorage = "true">属性才能编译成功。

9.4 实例工程：动态列表（GET 方式）

本实例演示了使用 HttpURLConnection 类发送 get 请求获取动态列表的数据，服务器返回动态列表的 JSON 数据，对 JSON 数据进行解析后，使用 ListView 控件显示动态的昵称和内容，下载图片并显示（如图 9-5 所示）。

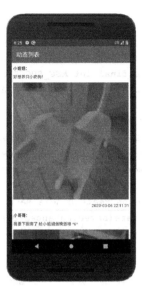

图 9-5　运行效果

1．打开基础工程

打开"基础工程"文件夹中的"C0906"工程，该工程已经包含 MainActivity、DailyAddActivity、DailyListActivity、DailyModel、DailyListViewAdapter、Http、ImageLoader、CompressImage、Util 类及布局的文件。

```
/java/com/vt/c0906/Http.java
148    //get 请求
149    public static String get(String requestUrl,String params) throws Exception{
150        URL url = new URL(requestUrl + params);
151        HttpURLConnection conn = (HttpURLConnection) url.openConnection();
152        conn.setConnectTimeout(CONNECT_TIME_OUT);
153        conn.setReadTimeout(READ_TIME_OUT);
154        conn.setRequestMethod("GET");
155        if (conn.getResponseCode() == 200) {
156            BufferedReader br= new BufferedReader(new InputStreamReader(conn.getInputStream()));
157            String state = br.readLine();//读取服务器端返回的数据
158            Log.e(TAG, "response getInputStream:" + state);
159            Log.e(TAG, "response Message:" + conn.getResponseMessage());
160            return state;
161        } else {
162            throw new RuntimeException(FAILURE);
163        }
164    }
```

第 150～154 行设置 get 请求的 HTTP 连接。第 155～164 行发送请求并读取响应码，当响应码等于 200 时，表示服务器响应成功，读取服务器返回的数据作为方法的返回值。

2．动态列表的 Activity

```
/java/com/vt/c0906/DailyListActivity.java
15    public class DailyListActivity extends AppCompatActivity {
```

```java
16     public static final String DOMAIN = "http://www.weiju2014.com/teachol/android/";
17     private String url = DOMAIN + "DailyList.php";
18     private static final int MSG_NETERROR = 0;
19     private static final int MSG_SUCCESS = 1;
20     private static final int MSG_NONDATA = 2;
21     private Context mContext = this;
22     private ListView mListView;
23     private DailyListViewAdapter mAdapter;
24     private DailyModel mDailyModel = new DailyModel();
25     private int mMaxAmount = 20;//设置动态最大显示数量
26     @Override
27     protected void onCreate(Bundle savedInstanceState) {
28         super.onCreate(savedInstanceState);
29         setContentView(R.layout.activity_daily_list);
30         this.setTitle("动态列表");
31         mContext = this;
32         mAdapter = new DailyListViewAdapter(DOMAIN,mContext, mDailyModel, R.layout.list_view_item_daily_list);
33         mListView = findViewById(R.id.list_view);
34         mListView.setAdapter(mAdapter);
35         loader(mMaxAmount,0);
36     }
37     //加载启动线程
38     public void loader(int n, int start) {
39         LoaderThread loaderThread = new LoaderThread();
40         loaderThread.n = n;
41         loaderThread.start = start;
42         loaderThread.start();
43     }
44     //加载线程
45     public class LoaderThread extends Thread {
46         private int n;
47         private int start;
48         @Override
49         public void run() {
50             addLists(n, start);
51         }
52     }
53     //获取加载数据
54     private int addLists(int n, int start) {
55         int num = 0;
56         try {
57             String res = Http.get(url, "?&num=" + n + "&start=" + start);//获取动态列表数据
58             //解析JSON数据
59             JSONTokener jsonParser = new JSONTokener(res.trim());
60             JSONObject person = (JSONObject) jsonParser.nextValue();
61             Log.e("json", person.getString("state"));
62             if (person.getString("state").equals("ok")) {
63                 JSONArray jsonObject = new JSONArray(person.getString("dailyList"));
64                 num = jsonObject.length();
```

```
65              for (int i = 0; i < jsonObject.length(); i++) {
66                  JSONObject jo = (JSONObject) jsonObject.opt(i);
67                  mDailyModel.ides.add(jo.getString("Id"));
68                  mDailyModel.names.add(jo.getString("Name"));
69                  mDailyModel.contents.add(jo.getString("Content"));
70                  mDailyModel.images.add(jo.getString("Image"));
71                  mDailyModel.createTimes.add(jo.getString("CreateTime"));
72              }
73              mHandler.obtainMessage(MSG_SUCCESS).sendToTarget();
74          } else {
75              mHandler.obtainMessage(MSG_NONDATA).sendToTarget();
76          }
77      } catch (Exception e) {
78          e.printStackTrace();
79          mHandler.obtainMessage(MSG_NETERROR).sendToTarget();
80      }
81      return num;
82  }
83  //更新UI
84  private Handler mHandler = new Handler() {
85      public void handleMessage(Message msg) {
86          switch (msg.what) {
87              case MSG_NETERROR://网络连接失败
88                  Toast.makeText(mContext, "网络连接失败", Toast.LENGTH_LONG).show();
89                  break;
90              case MSG_SUCCESS://加载
91                  mAdapter.notifyDataSetChanged();
92                  break;
93              case MSG_NONDATA://没有
94                  break;
95          }
96      }
97  };
98  }
```

第 38~43 行开启独立线程加载动态列表的数据。第 45~52 行 LoaderThread 类继承自 Thread 类，调用 addLists(int, int)方法加载数据。第 54~82 行通过 Http.get(String,String)方法获取服务器返回的 JSON 数据。如果 JSON 数据中的 state 元素值为 ok，则对动态列表数据进行解析，并保存到 mDailyModel 对象，然后通过 mHandler.obtainMessage(MSG_SUCCESS).sendToTarget()发送加载成功的消息，否则发送加载失败的消息。第 84~97 行通过 handleMessage(Message)方法处理发送过来的消息，当接收到 MSG_SUCCESS 消息时，调用 mAdapter.notifyDataSetChanged()方法更新 ListView 控件显示的动态列表。

3．测试运行

运行 C0906 工程后，单击"动态列表"按钮启动 DailyListActivity。列表数据加载完成后，在"Logcat"窗口中可以查看输出的 JSON 数据。

图 9-6 "Logcat" 窗口输出的数据

扩展实例：下拉刷新和加载更多的 ListView

在微博、小红书、哔哩哔哩等 App 中，可以通过向下滑动 ListView 控件实现刷新的效果，而且滑动到 ListView 控件的底部时可以自动继续加载列表数据或手动单击"更多"按钮加载列表数据。实现这些功能的代码较多，由于篇幅所限，C0907 工程提供了这些功能（如图 9-7 所示），可自行查看代码。

图 9-7 带下拉刷新和更多加载的 ListView 控件

9.5 习　　题

1. 实现文件下载的功能，将文件保存在缓存文件夹中。

2. 实现服务器端登录验证的功能，验证地址为 http://www.weiju2014.com/teachol/android/login.php，使用 get 方法，用户名的参数为 name，密码的参数为 password。可以使用的登录用户名为 abc，密码为 000，登录成功后返回的 JSON 数据为{"state":"ok"}，登录失败则返回{"state":"error"}。

3. 实现服务器端获取动态中的图片并通过 ListView 控件显示图片，获取最近发布的 100 条动态数据，地址为 http://www.weiju2014.com/teachol/android/DailyList.php?num= 100&start=0。